# The determination and interpretation of molecular wave functions

ERICH STEINER

*Lecturer in Chemistry, University of Exeter*

T0291445

**Cambridge University Press**

CAMBRIDGE

LONDON · NEW YORK · MELBOURNE

CAMBRIDGE UNIVERSITY PRESS
Cambridge, New York, Melbourne, Madrid, Cape Town, Singapore, São Paulo, Delhi

Cambridge University Press
The Edinburgh Building, Cambridge CB2 8RU, UK

Published in the United States of America by Cambridge University Press, New York

www.cambridge.org
Information on this title: www.cambridge.org/9780521105675

First published 1976
This digitally printed version 2009

A catalogue record for this publication is available from the British Library

Library of Congress Catalogue Card Number: 75–18120

ISBN 978-0-521-21037-9 hardback
ISBN 978-0-521-10567-5 paperback

# Preface

It is a fact of some historical interest that although a comprehensive and detailed theory of the electronic structure and bonding in molecules had been developed by the year 1960 (Pauling 1960; Coulson 1961), very few non-empirical ('*ab initio*') quantum-mechanical calculations for polyatomic molecules were performed before that date. The theory was built up almost wholly intuitively and empirically from an experimental knowledge of the physical and chemical properties of molecules, coupled with an extrapolation to polyatomic molecules of the results of (i) accurate solutions of the Schrödinger equation for the hydrogen molecule and molecular ion, (ii) highly approximate non-empirical calculations for small, and mainly diatomic, molecules, and (iii) semi-empirical calculations, mainly of the Hückel molecular-orbital type, for some larger molecules. It was not until the late 1950s that the advent of the electronic computer brought with it the possibility of performing accurate non-empirical calculations for polyatomic molecules of chemical interest. Much of the subsequent evolution of the theory of molecular structure has paralleled, and has to a certain extent depended on, the development of computing machines and techniques.

The advent of the computer has led to the creation of a new branch of theoretical chemistry, computational quantum chemistry, with its own specialized language, and with concepts that are increasingly influenced by questions of mathematical tractability and computational expedience. One consequence has been the creation of new problems of communication between theoretical chemist and experimental chemist, and this book can be regarded as an attempt to bridge the widening gap between the two. The primary concern of the book is the exposition of some of the more important theoretical and computational techniques that have been developed in recent years for the determination and interpretation of molecular wave functions, with particular emphasis on the non-empirical molecular-orbital approach. A feature of the evolution of the theory since 1960 has been the declining importance of the valence-bond approach as a practical tool of the computational quantum chemist. Although valence-bond theory is as valid as molecular-orbital theory, and merely represents an alternative method of constructing molecular wave functions, molecular-orbital theory has been found to be the more con-

# Preface

venient for computational purposes and, to a lesser extent, for interpretative purposes.

It has been assumed that the reader is a graduate or advanced undergraduate in chemistry, with the appropriate knowledge of mathematics and of the applications of quantum mechanics in chemistry. Chapter 1 is devoted to a brief discussion of the non-relativistic time-independent Schrödinger equation for the motion of electrons in molecules, and of the general techniques available for its solution. Chapter 2 deals with the symmetry properties of electronic wave functions. Nearly all of the material in these first two chapters can be found discussed at greater length in standard undergraduate texts, but it has been included both to make the book more or less self-contained and to introduce the notation, units, and those general concepts that are used in the subsequent discussion. Chapter 3 is devoted to the Hartree–Fock model of electronic structure, and to its relation to what is commonly known as molecular-orbital theory. Methods of proceeding beyond the orbital approximation towards the exact solution of the Schrödinger equation are considered in chapter 4, which also includes a brief discussion of relativistic effects. Chapter 5 is concerned with some of the computational techniques that have been developed for the practical implementation of the theory developed in the previous chapters. The analysis and interpretation of molecular wave functions is discussed in the final two chapters. Chapter 6 is concerned with the electron distribution and chapter 7 with the nature of the chemical bond.

I wish to acknowledge with gratitude the encouragement and advice given to me by several of my colleagues in the Chemistry Department. Particularly I would like to thank Dr B. J. Skillerne de Bristowe for reading much of the manuscript during the earlier stages of preparation, and for many helpful suggestions.

E. S.

*University of Exeter*
*May 1975*

# 1 Introduction

## 1.1 THE SCHRÖDINGER EQUATION

We are concerned in this book with the non-empirical theory of the electronic structure of molecules. By a non-empirical (or *ab initio*) calculation in molecular quantum chemistry is normally meant the solution of a time-independent Schrödinger equation

$$\mathcal{H}\Psi = E\Psi \qquad (1.1)$$

in which the Hamiltonian $\mathcal{H}$ is that appropriate to any model of the system which does not depend, either explicitly or implicitly, on the properties of any finite number of states of the system. In the simplest and most widely used model of a molecule, the nuclei and electrons are assumed to be non-relativistic point charges interacting through electrostatic (Coulomb) forces only, the force acting between charges $q_1$ and $q_2$ separated by distance $r$ being $q_1 q_2/4\pi\epsilon_0 r^2$, where $\epsilon_0$ is the permittivity of a vacuum.

Consider a molecule containing $\nu$ nuclei, with charges $Z_\alpha e$ and masses $M_\alpha$ ($\alpha = 1, 2, ..., \nu$), and $N$ electrons, with charges $-e$ and masses $m_e$. Let the position of nucleus $\alpha$ be given by the vector $\boldsymbol{R}_\alpha$, whose components are the coordinates of the nucleus in a fixed coordinate system, and let $\boldsymbol{r}_i$ be the position vector of electron $i$ ($i = 1, 2, ..., N$). The Hamiltonian for this system of point charges is

$$\mathcal{H} = \sum_{\alpha=1}^{\nu} \frac{-h^2}{8\pi^2 M_\alpha} \nabla_\alpha^2 + \sum_{i=1}^{N} \frac{-h^2}{8\pi^2 m_e} \nabla_i^2$$

$$- \sum_{i=1}^{N}\sum_{\alpha=1}^{\nu} \frac{Z_\alpha e^2}{4\pi\epsilon_0 r_{i\alpha}} + \sum_{i>j=1}^{N}\sum \frac{e^2}{4\pi\epsilon_0 r_{ij}} + \sum_{\alpha>\beta=1}^{\nu}\sum \frac{Z_\alpha Z_\beta e^2}{4\pi\epsilon_0 R_{\alpha\beta}} \qquad (1.2)$$

where $h$ is Planck's constant and, for example, $r_{i\alpha} = |\boldsymbol{R}_\alpha - \boldsymbol{r}_i|$ is the distance between electron $i$ and nucleus $\alpha$. The corresponding Schrödinger equation for this model has been found, through many applications, to form a satisfactory basis for the description of a very wide variety of properties of molecules, and we will be concerned in this book almost wholly with the methods that have been developed for its solution and with the interpretation of the solutions. Apart from a brief discussion

I

in §4.5 of the magnitudes of the relativistic corrections to the model, we will therefore consider as outside the scope of this book all time-dependent phenomena, the effects of external fields, as well as all magnetic interactions and other relativistic effects. Inclusion of these requires modification either of the Hamiltonian or of the form of the wave equation itself, although these modifications are often treated as perturbations on the model system (§1.5).

The Hamiltonian (1.2) contains terms which describe not only the motion of the electrons about the nuclei, but also the motion of the nuclei with respect to each other and of the molecule as a whole in space. The corresponding Schrödinger equation is very difficult to solve in general, and accurate solutions have been obtained only for the simplest molecules, $H_2^+$ and $H_2$ (Kolos and Wolniewicz 1964). The problem is simplified considerably however by including in the model further assumptions which result in the separation of the electronic and nuclear motions. These assumptions rely on the observation that the ratio $m_e/M_\alpha$ of the electronic to nuclear masses is a small number compared with unity, and they are therefore consistent with our definition of a non-empirical theory. The nuclear and electronic motions may be separated exactly for a one-electron atom, and to a very good approximation for other atoms (Bethe and Salpeter 1957). The resulting Hamiltonian for the internal motion of an atom, whose nucleus has charge $Ze$ and mass $M$, is

$$\mathscr{H} = \frac{-h^2}{8\pi^2\mu} \sum_{i=1}^{N} \nabla_i^2 - \sum_{i=1}^{N} \frac{Ze^2}{4\pi\epsilon_0 r_i} + \sum_{i>j=1}^{N} \frac{e^2}{4\pi\epsilon_0 r_{ij}} \qquad (1.3)$$

where $\mu = m_e M/(m_e + M)$ is the reduced mass of an electron in the atom, and $r_i$ is the position vector of electron $i$ relative to the nucleus as origin. The separation for molecules is of a somewhat different kind, the Hamiltonian for electronic motion being obtained by assuming that the nuclei have infinite masses and, therefore, have fixed positions relative to a fixed coordinate system. The Hamiltonian for the motion of the electrons in this 'fixed-nuclei' or Born–Oppenheimer approximation is

$$\mathscr{H} = \frac{-h^2}{8\pi^2 m_e} \sum_{i=1}^{N} \nabla_i^2 - \sum_{i=1}^{N} \sum_{\alpha=1}^{\nu} \frac{Z_\alpha e^2}{4\pi\epsilon_0 r_{i\alpha}}$$

$$+ \sum_{i>j=1}^{N} \frac{e^2}{4\pi\epsilon_0 r_{ij}} + \sum_{\alpha>\beta=1}^{\nu} \frac{Z_\alpha Z_\beta e^2}{4\pi\epsilon_0 R_{\alpha\beta}} \qquad (1.4)$$

The solutions of the corresponding Schrödinger equation depend on the nuclear positions, and a separate calculation of any electronic state must be performed for each assumed molecular geometry. The stable geometry (in the Born–Oppenheimer approximation) for any state is that with the lowest energy.

The eigenfunctions of the Hamiltonian, (1.3) for an atom and (1.4) for a molecule, describe the stationary electronic states of the system. They are functions of the coordinates of the electrons, and they can always be chosen to be normalized and orthogonal (orthonormal); if $\Psi_m$ and $\Psi_n$ are any two eigenfunctions,

$$\int \Psi_m^* \Psi_n \, d\tau = \begin{cases} 1 & \text{if } m = n, \quad \text{for normalization} \\ 0 & \text{if } m \neq n, \quad \text{for orthogonality} \end{cases}$$

where $\Psi_m^*$ is the complex conjugate of $\Psi_m$, and $\int \ldots d\tau$ implies integration over all the coordinates of the electrons. An important property of the eigenfunctions is that they form a *complete set* of functions in the sense that any arbitrary wave function $\Psi$, which is not an eigenfunction of $\mathscr{H}$ but which satisfies the same boundary conditions as the eigenfunctions, can be expressed as a linear combination of the eigenfunctions:

$$\Psi = \sum_n C_n \Psi_n$$

If the eigenfunctions are orthonormal, the coefficients are given by

$$C_n = \int \Psi_n^* \Psi \, d\tau$$

## 1.2 ATOMIC UNITS

The Schrödinger equation with Hamiltonian (1.4) for a Born–Oppenheimer molecule may be freed of the experimentally determined quantities $e$, $h$, $\epsilon_0$ and $m_e$ by the substitutions

$$\mathscr{H} = (m_e e^4 / 4h^2 \epsilon_0^2) \mathscr{H}', \quad E = (m_e e^4 / 4h^2 \epsilon_0^2) E'$$

and, for example,

$$r_{ij} = (\epsilon_0 h^2 / \pi m_e e^2) r_{ij}'$$

The conversion factors are often treated as units, atomic units (Shull and Hall 1959). They are the Bohr radius $a_\infty$ (or $a_0$) and the Hartree energy $H_\infty$:

$$\left. \begin{aligned} a_\infty &= \epsilon_0 h^2 / \pi m_e e^2 = 5.2918 \times 10^{-11} \, \text{m} \\ H_\infty &= m_e e^4 / 4h^2 \epsilon_0^2 = e^2 / 4\pi \epsilon_0 a_\infty = 4.3598 \times 10^{-18} \, \text{J} \end{aligned} \right\} \quad (1.5)$$

3

# Introduction

The resulting dimensionless Schrödinger equation and Hamiltonian are

$$\mathscr{H}'\Psi' = E'\Psi'$$

$$\left. \mathscr{H}' = -\tfrac{1}{2}\sum_{i=1}^{N}\nabla_i'^2 - \sum_{i=1}^{N}\sum_{\alpha=1}^{\nu}\frac{Z_\alpha}{r_{i\alpha}'} + \sum_{i>j=1}^{N}\frac{1}{r_{ij}'} + \sum_{\alpha>\beta=1}^{\nu}\frac{Z_\alpha Z_\beta}{R_{\alpha\beta}'} \right\} \quad (1.6)$$

and the primes, which convert energies and lengths to numbers, are in practice omitted.

The use of the symbols $a_\infty$ and $H_\infty$ for the atomic units of length and energy requires a few words of explanation. The most commonly used symbol for the Bohr radius is $a_0$, but no generally accepted symbol exists for the Hartree energy, and many authors avoid the use of special symbols by denoting the atomic unit of every physical quantity by the abbreviation a.u. The symbols $a_\infty$ and $H_\infty$ proposed here have been chosen to be consistent with the SI-recommended symbol $R_\infty$ for the Rydberg constant, which is related to the Hartree energy by $H_\infty = 2hcR_\infty$, where $c$ is the speed of light. In the case of an atom whose Hamiltonian (1.3) involves the reduced mass $\mu$ instead of the electronic mass $m_e$, the Schrödinger equation is reduced to the dimensionless form (1.6), not involving $\mu$, by the conversion factors obtained from $a_\infty$ and $H_\infty$ by replacing $m_e$ by $\mu$. The new conversion factors can be distinguished from $a_\infty$ and $H_\infty$ by a change of subscript to specify the nature of the nucleus. Thus the Rydberg constant for the normal ($^1$H) hydrogen atom is

$$R_H = R_\infty m_p/(m_p + m_e)$$

where $m_p$ is the mass of the proton, and the corresponding symbols for the length and energy are $a_H$ and $H_H$. In general for any subscript X,

$$H_X = 2hcR_X \quad \text{and} \quad a_X H_X = e^2/4\pi\epsilon_0$$

In this book we shall ignore, for simplicity, the small difference in value between $m_e$ and the reduced mass $\mu$ of an electron in an atom; this corresponds to the Born–Oppenheimer assumption of infinite nuclear mass.

Any four of the five quantities $m_e$, $e$, $\hbar = h/2\pi$, $a_\infty$ and $H_\infty$ may be regarded as base atomic units for the construction of the atomic units of all those other physical quantities which, in SI, involve only the units of length mass, time and electric current. A list of some of the more important quantities is given in table 1.1. Other quantities used as units in this book include the ångström $\text{Å} = 10^{-10}\,\text{m} = 1.8897 a_\infty$; the debye $D = 3.3356 \times 10^{-30}\,\text{Cm}$, which is related to the atomic unit of electric dipole moment by $e a_\infty = 2.5418\,D$; the electron volt

4

TABLE 1.1  *Atomic units*

| Physical quantity | Atomic unit | Value in SI units |
| --- | --- | --- |
| Mass | $m_e$ | $9.1096 \times 10^{-31}$ kg |
| Charge | $e$ | $1.6022 \times 10^{-19}$ C |
| Angular momentum | $\hbar = h/2\pi$ | $1.0546 \times 10^{-34}$ J s |
| Length | $a_\infty = 4\pi\epsilon_0\hbar^2/m_e e^2$ | $5.2918 \times 10^{-11}$ m |
| Energy | $H_\infty = m_e e^4/16\pi^2\epsilon_0^2\hbar^2$ | $4.3598 \times 10^{-18}$ J |
| Time | $\hbar/H_\infty$ | $2.4189 \times 10^{-17}$ s |
| Linear momentum | $\hbar/a_\infty$ | $1.9928 \times 10^{-24}$ kg m s$^{-2}$ |
| Electric current | $eH_\infty/\hbar$ | $6.6237 \times 10^{-3}$ A |
| Electric potential | $H_\infty/e$ | $2.7211 \times 10^1$ V |
| Electric dipole moment | $ea_\infty$ | $8.4784 \times 10^{-30}$ C m |
| Electric charge density | $e/a_\infty^3$ | $1.0812 \times 10^{12}$ C m$^{-3}$ |

eV $= 1.6022 \times 10^{-19}$ J, with $H_\infty = 27.211$ eV; and the molar energy $LH_\infty = 2.6255 \times 10^6$ J mol$^{-1}$, where $L$ is the Avogadro constant.

### 1.3  THE VARIATION PRINCIPLE

In a non-empirical calculation one attempts to find eigenfunctions and eigenvalues of the 'exact' model Hamiltonian (1.6). Except for the simplest systems however, a complete solution of the Schrödinger equation is still an intractable problem, and it is therefore always necessary to resort to methods of finding approximate solutions. Almost all of these methods are based on the variation principle.

A simple expression of the variation principle is that, given any trial $N$-electron wave function $\Psi$ which satisfies the necessary boundary conditions for the system, an upper bound to the exact ground-state energy $E_0$ is

$$E = \frac{\int \Psi^* \mathscr{H} \Psi \, d\tau}{\int \Psi^* \Psi \, d\tau} \geq E_0 \qquad (1.7)$$

Analogous inequalities exist for excited states. A consequence of the principle is that if a trial wave function depends on a number of arbitrary parameters, $\lambda_1, \lambda_2, ..., \lambda_n$,

$$\Psi = \Psi(\mathbf{r}; \lambda_1, \lambda_2, ..., \lambda_n)$$

where $\mathbf{r}$ represents the dependence of $\Psi$ on the coordinates of the electrons, then the values of these parameters can be chosen to give the lowest possible, and hence the most accurate, value of the energy. The energy is a

5

function of the parameters, and the values of the parameters which give the lowest value of the energy are obtained by solving the equations

$$\frac{\partial E}{\partial \lambda_i} = 0 \quad (i = 1, 2, ..., n)$$

The most general type of approximate wave function commonly used has the form

$$\Psi = C_1 \Phi_1 + C_2 \Phi_2 + ... + C_n \Phi_n = \sum_{j=1}^{n} C_j \Phi_j \qquad (1.8)$$

where the coefficients $C_j$ are parameters, and the $\Phi_j$ are given $N$-electron functions which satisfy the same boundary conditions as $\Psi$ and which may or may not depend on further parameters. The corresponding energy (1.7) is

$$E = \sum_{i, j=1}^{n} C_i^* C_j H_{ij} \Big/ \sum_{i, j=1}^{n} C_i^* C_j S_{ij} \qquad (1.9)$$

where

$$H_{ij} = \int \Phi_i^* \mathscr{H} \Phi_j \, d\tau, \quad S_{ij} = \int \Phi_i^* \Phi_j \, d\tau \qquad (1.10)$$

and the minimization of $E$ with respect to the $n$ coefficients gives a set of $n$ 'secular' equations

$$\sum_{j=1}^{n} (H_{ij} - ES_{ij}) C_j = 0 \quad (i = 1, 2, ..., n) \qquad (1.11)$$

One (trivial) solution of the equations is obtained by setting all the coefficients equal to zero. Non-trivial solutions are obtained only if the energy $E$ is chosen such that the secular determinant, whose elements are $(H_{ij} - ES_{ij})$, vanishes: $\det(H_{ij} - ES_{ij}) = 0$

or

$$\begin{vmatrix} H_{11} - ES_{11} & H_{12} - ES_{12} & \cdots & H_{1n} - ES_{1n} \\ H_{21} - ES_{21} & H_{22} - ES_{22} & \cdots & H_{2n} - ES_{2n} \\ \cdots\cdots\cdots\cdots\cdots\cdots\cdots\cdots\cdots\cdots\cdots\cdots\cdots\cdots \\ H_{n1} - ES_{n1} & H_{n2} - ES_{n2} & \cdots & H_{nn} - ES_{nn} \end{vmatrix} = 0 \quad (1.12)$$

The secular determinant is a polynomial of degree $n$ in the energy, and it has $n$ roots, not necessarily all different,

$$E_1 \leqslant E_2 \leqslant E_3 \leqslant ... \leqslant E_n$$

Corresponding to each energy $E_i$, a wave function

$$\Psi_i = \sum_{j=1}^{n} \Phi_j C_{ji} \qquad (1.13)$$

may now be obtained by solving the secular equations and normalization. The resulting wave functions are orthonormal:

$$\int \Psi_i^* \Psi_j \, d\tau = \sum_{k,\,l=1}^{n} C_{ki}^* C_{lj} S_{kl} = \delta_{ij} = \begin{cases} 1 & \text{if} \quad i = j \\ 0 & \text{if} \quad i \neq j \end{cases}$$

The lowest root $E_1$ is an approximate ground-state energy, and the corresponding function $\Psi_1$ is an approximate wave function for the ground state. In fact, the set of $n$ solutions are approximations for the first $n$ states of the system. If $E_i^{(e)}$ is the exact energy of the $i$th state, then $E_i \geqslant E_i^{(e)}$ as shown in fig 1.1, and $E_i = E_i^{(e)}$ only if $\Psi_i$ is the exact wave function for the $i$th state. The magnitudes of the separations $(E_i - E_i^{(e)})$ depend on the functions $\Phi_j$ included in the wave function (1.8). If the functions depend on further parameters,

$$\Phi_j = \Phi_j(\boldsymbol{r}; \lambda_1, \lambda_2, \ldots)$$

then the parameters can be chosen to minimize one of the roots $E_i$ of the secular problem.

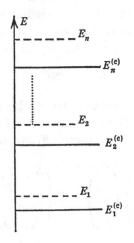

Fig. 1.1

It is generally true that increasing the number of variational parameters in a wave function results in improved accuracy of the corresponding energy. If the wave function has a general enough form, the exact solution is approached as the number of parameters is increased indefinitely. In particular, the 'method of linear combinations' outlined above provides better approximations to more and more states as the number $n$ increases. In practice however the form of the wave function is often constrained, and cannot lead to an exact solution of the Schrödinger equation. An example is the orbital approximation, discussed in some detail in chapter 3, in which the wave function and energy approach definite (Hartree–Fock) limiting values as the number of variational parameters increases, but these limits do not represent an exact solution of the Schrödinger equation. As we shall see, the wave function obtained in this way is an eigenfunction not of the Hamiltonian (1.6) but of a different Hamiltonian, the Hartree–Fock Hamiltonian. In this way a new model based on the *form* of the wave function is obtained as an approximation to the original 'exact' model, to which it reduces when the constraints imposed on the form of the wave function are relaxed.

## 1.4 MATRIX REPRESENTATION OF THE SCHRÖDINGER EQUATION

The Schrödinger equation for an $N$-electron system is a partial differential equation in $3N$ variables, but the method of linear combinations discussed in the previous section shows how it may be transformed into an equivalent matrix equation. The secular equations (1.11) can be written in the matrix form

$$HC = ESC \tag{1.14}$$

where $H$ and $S$ are square ($n \times n$) matrices whose elements are $H_{ij}$ and $S_{ij}$ defined by (1.10), and $C$ is a column matrix (vector) whose elements are the coefficients $C_j$ of the expansion (1.8); written out in full, (1.14) is

$$
\begin{pmatrix}
H_{11} & H_{12} & \cdots & H_{1n} \\
H_{21} & H_{22} & \cdots & H_{2n} \\
\cdots & & & \\
H_{n1} & H_{n2} & \cdots & H_{nn}
\end{pmatrix}
\begin{pmatrix}
C_1 \\
C_2 \\
\cdots \\
C_n
\end{pmatrix}
= E
\begin{pmatrix}
S_{11} & S_{12} & \cdots & S_{1n} \\
S_{21} & S_{22} & \cdots & S_{2n} \\
\cdots & & & \\
S_{n1} & S_{n2} & \cdots & S_{nn}
\end{pmatrix}
\begin{pmatrix}
C_1 \\
C_2 \\
\cdots \\
C_n
\end{pmatrix}
$$

The matrix $H$ is a representation of the Hamiltonian $\mathscr{H}$ in terms of the *basis* of $n$ functions $\Phi_j$. As has already been remarked, the solutions of the matrix equation are approximations for $n$ states of the system, and increasing the number of basis functions leads to better approximations for more and more states. When $n$ becomes infinitely large and the basis becomes complete,† the matrix equation (1.14) becomes entirely equiva-

---

† The definition of a complete set given on p. 3 is sufficient for our purposes. Given a set of functions $\Phi_n$ ($n = 1, 2, ..., M$) satisfying certain boundary conditions, the set is said to be complete if any arbitrary function $\Phi$, which satisfies the same boundary

lent to the differential Schrödinger equation, having the same set of eigenvalues and eigenfunctions, the latter being expressed in the form (1.13) as linear combinations of the basis functions.

Given a basis therefore, the problem of solving the Schrödinger equation is reduced to the evaluation of the matrix elements $H_{ij}$ and $S_{ij}$, and to the solution of the corresponding matrix equation. The only serious problem is concerned with the choice of basis functions. Ideally, we would like to be able to use a basis which gives both easily evaluated matrix elements and a rapid convergence (small $n$) of the expansion (1.13) of the wave function. This is not always possible, particularly for polyatomic molecules.

### 1.5 PERTURBATION THEORY

It is often the case that the Hamiltonian $\mathscr{H}$ for the system of interest differs only slightly from the Hamiltonian $\mathscr{H}_0$ of a related system. One example is a molecule in a weak external electric field for which the Hamiltonian can be written as

$$\mathscr{H} = \mathscr{H}_0 + \lambda V \tag{1.15}$$

where $\mathscr{H}_0$ describes the unperturbed system which is the molecule in the absence of the field, and $\lambda V$ is a 'small' perturbation term which describes the interaction of the molecule with the field. $\lambda$ is a parameter, called the perturbation parameter (for example, the field strength), which is a measure of the strength of the perturbation.

It is generally assumed in perturbation theory that the eigenfunctions and eigenvalues of $\mathscr{H}_0$ are known,

$$\mathscr{H}_0 \Psi_n^{(0)} = E_n^{(0)} \Psi_n^{(0)} \tag{1.16}$$

or, since in practice only a few eigenfunctions may be known, that at least the unperturbed wave function $\Psi_n^{(0)}$ for the state of interest is known. We shall also assume, for simplicity, that the state of interest is one for which the energy $E_n^{(0)}$ is non-degenerate. Then, if $\Psi_n$ and $E_n$ are the

conditions, can be expressed as a linear combination $\Phi = \sum_{n=1}^{M} C_n \Phi_n$. If the basis functions are orthonormal, $\int \Phi_m^* \Phi_n \, d\tau = \delta_{mn}$, the coefficients are given by
$$C_n = \int \Phi_n^* \Phi \, d\tau,$$
Such complete sets of functions are almost always infinite, notable exceptions being the sets of $N$-electron spin functions containing $2^N$ members (§2.7).

9

*Introduction*

(unknown) eigenfunction and eigenvalue of $\mathcal{H}$ for this state,

$$\mathcal{H}\Psi_n = E_n\Psi_n \tag{1.17}$$

it follows that in the limit $\lambda \to 0$,

$$\mathcal{H} \to \mathcal{H}_0, \quad E_n \to E_n^{(0)}, \quad \Psi_n \to \Psi_n^{(0)}$$

The basic assumption of perturbation theory is that the energy and wave function for the perturbed state may be expanded as power series in $\lambda$ about the corresponding energy and wave function of the unperturbed state,

$$\left.\begin{aligned} E_n &= E_n^{(0)} + \lambda E_n^{(1)} + \lambda^2 E_n^{(2)} + \dots \\ \Psi_n &= \Psi_n^{(0)} + \lambda \Psi_n^{(1)} + \lambda^2 \Psi_n^{(2)} + \dots \end{aligned}\right\} \tag{1.18}$$

and that these expansions are valid for the whole range of values of $\lambda$ between zero and the value of interest. The quantity $E_n^{(i)}$ is called the $i$th-order energy and $\Psi_n^{(i)}$ is the $i$th-order wave function. Substitution of the expansions for the energy and wave function in the Schrödinger equation (1.17) gives

$$(\mathcal{H}_0 + \lambda V)(\Psi_n^{(0)} + \lambda \Psi_n^{(1)} + \lambda^2 \Psi_n^{(2)} + \dots)$$
$$= (E_n^{(0)} + \lambda E_n^{(1)} + \lambda^2 E_n^{(2)} + \dots)(\Psi_n^{(0)} + \lambda \Psi_n^{(1)} + \lambda^2 \Psi_n^{(2)} + \dots)$$

or

$$(\mathcal{H}_0 - E_n^{(0)})\Psi_n^{(0)} + \lambda[(\mathcal{H}_0 - E_n^{(0)})\Psi_n^{(1)} + (V - E_n^{(1)})\Psi_n^{(0)}]$$
$$+ \lambda^2[(\mathcal{H}_0 - E_n^{(0)})\Psi_n^{(2)} + (V - E_n^{(1)})\Psi_n^{(1)} - E_n^{(2)}\Psi_n^{(0)}] + \dots = 0 \tag{1.19}$$

In order that the equation be satisfied for arbitrary values of $\lambda$, it is necessary that the coefficient of each power of $\lambda$ be separately zero. The zeroth-order equation is

$$(\mathcal{H}_0 - E_n^{(0)})\Psi_n^{(0)} = 0$$

which is simply the original Schrödinger equation for the unperturbed state. The first- and second-order equations are

$$(\mathcal{H}_0 - E_n^{(0)})\Psi_n^{(1)} + (V - E_n^{(1)})\Psi_n^{(0)} = 0 \tag{1.20}$$

$$(\mathcal{H}_0 - E_n^{(0)})\Psi_n^{(2)} + (V - E_n^{(1)})\Psi_n^{(1)} = E_n^{(2)}\Psi_n^{(0)} \tag{1.21}$$

We therefore obtain, if the perturbation $\lambda V$ is small enough, a set of equations which can be solved in sequence to give progressively more accurate solutions of the Schrödinger equation for the perturbed state. In many applications it is not necessary to go beyond the first-order wave function as this determines the energy to third order.

The solution of the perturbation equations is discussed in standard

10

texts, and only the more important results are summarized here. The energies are

$$E_n^{(1)} = \int \Psi_n^{(0)*} V \Psi_n^{(0)} \, d\tau \qquad (1.22)$$

$$E_n^{(2)} = \int \Psi_n^{(0)*} V \Psi_n^{(1)} \, d\tau \qquad (1.23)$$

and, in general, $\quad E_n^{(i)} = \int \Psi_n^{(0)*} V \Psi_n^{(i-1)} \, d\tau$

We note that the energy to first order,

$$E_n^{(0)} + \lambda E_n^{(1)} = \int \Psi_n^{(0)*} (\mathscr{H}_0 + \lambda V) \Psi_n^{(0)} \, d\tau = \int \Psi_n^{(0)*} \mathscr{H} \Psi_n^{(0)} \, d\tau$$

is simply the energy of the perturbed state in terms of the approximate wave function $\Psi_n^{(0)}$. Alternative expressions for the second- and third-order energies are

$$E_n^{(2)} = \int \Psi_n^{(1)*} (\mathscr{H}_0 - E_n^{(0)}) \Psi_n^{(1)} \, d\tau + 2 \int \Psi_n^{(0)*} V \Psi_n^{(1)} \, d\tau \qquad (1.24)$$

$$E_n^{(3)} = \int \Psi_n^{(1)*} (V - E_n^{(1)}) \Psi_n^{(1)} \, d\tau \qquad (1.25)$$

and the latter shows that the first-order wave function determines the energy to third order.

The eigenfunctions $\Psi_n^{(0)}$ of the Hamiltonian $\mathscr{H}_0$ of the unperturbed system form a complete orthonormal set, and the first-order wave function may therefore be expressed as a linear combination of them. A formal solution of the first-order equation (1.20) of this form is

$$\Psi_n^{(1)} = \sum_{m \neq n} \left( \frac{V_{mn}}{E_n^{(0)} - E_m^{(0)}} \right) \Psi_m^{(0)} \qquad (1.26)$$

where† $\qquad V_{mn} = \int \Psi_m^{(0)*} V \Psi_n^{(0)} \, d\tau$

The corresponding expression for the second-order energy is

$$E_n^{(2)} = \sum_{m \neq n} \frac{V_{nm} V_{mn}}{E_n^{(0)} - E_m^{(0)}} \qquad (1.27)$$

Evaluation of the sum requires a knowledge of the complete set of unperturbed wave functions, and an alternative approach is to calculate

† Since $\int \Psi_m^{(0)*} \mathscr{H}_0 \Psi_n^{(0)} \, d\tau = 0$ if $m \neq n$, we can also write $\lambda V_{mn} = \int \Psi_m^{(0)*} \mathscr{H} \Psi_n^{(0)} \, d\tau$.

# Introduction

the first-order wave function by a variational method. It can be shown that the expression (1.24) gives an upper bound to the exact second-order energy when $\Psi_n'^{(1)}$ is replaced by an approximate function, $\Phi$ say:

$$E_n^{(2)} \leqslant W = \int \Phi^* (\mathscr{H}_0 - E_n^{(0)}) \Phi \, d\tau + 2 \int \Psi_n^{(0)*} V \Phi \, d\tau \qquad (1.28)$$

The approximate second-order energy $W$ may be minimized with respect to variations of $\Phi$, and this procedure has the advantage that only the unperturbed wave function $\Psi_n^{(0)}$ for the state in question need be known. The calculated approximate first-order wave function may then be used to calculate an approximate value of the third-order energy by means of (1.25).

# 2 Symmetry

## 2.1 INTRODUCTION

The exact solutions of the Schrödinger equation of an atom or molecule have certain symmetry properties which are a necessary consequence of the symmetry properties of the system. Although it is possible to obtain only approximate solutions in the general case, these symmetry properties are usually, and relatively simply, built into an approximate wave function by, for example, the use of projection operators (§2.4).

The symmetry properties of an atom or molecule are associated with the invariance of the Hamiltonian with respect to three types of operations:

(i) *Electron permutation symmetry*. A permutation of the coordinates of the (identical) electrons interchanges identical terms in the Hamiltonian, and therefore leaves the Hamiltonian unchanged (§2.5).

(ii) *Spatial symmetry*. The Hamiltonian of a molecule is invariant with respect to a spatial operation, such as a rotation about an axis of symmetry or a reflection through a plane of symmetry, which results in the interchange of identical nuclei. For a linear molecule, the cylindrical symmetry about the molecular axis results in the quantization of the component of the total angular momentum of the electrons along this axis. For an atom, the spherical symmetry results in the quantization both of the square of the angular momentum and of a component (§2.6).

(iii) *Spin symmetry*. The Hamiltonian does not contain spin-dependent terms and is therefore invariant with respect to the spin angular momentum operators. This results in the quantization of the square of the total spin angular momentum of the electrons and of a component along a specified direction (§2.7).

The mathematical theory of symmetry is group theory, and we begin this chapter with a brief discussion of the principles of group theory. More complete discussions and proofs of the theorems may be found in a number of excellent texts; for example, Cotton (1963), Bishop (1973), Eyring, Walter and Kimball (1944, chapter 10), and Wigner (1959).

## 2.2 GROUP THEORY

Consider the symmetrical plane figure formed by three points at the corners of an equilateral triangle (fig. 2.1). Let the figure lie in the $xy$

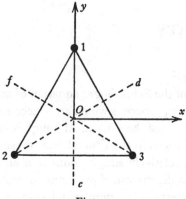

Fig. 2.1

plane of a fixed coordinate system whose origin $O$ is at the centroid of the triangle.

The symmetry properties of the figure can be described in terms of the following set of six symmetry operations which send the figure into itself:

$E$, the identity operation which leaves every point unmoved

$A$, anti-clockwise rotation through 120° about the $Oz$ axis

$B$, anti-clockwise rotation through 240° (or clockwise through 120°) about the $Oz$ axis

$C$, rotation through 180° about the $Oc$ axis

$D$, rotation through 180° about the $Od$ axis

$F$, rotation through 180° about the $Of$ axis

Other symmetry operations (reflections) are possible, but are equivalent to the above six operations. The operations may also be interpreted simply as permutations of the labels (1, 2, 3) of the three points.

The successive application of any two symmetry operations is equivalent to the application of a single operation. The two examples in fig. 2.2 show that the application of operation $A$ followed by $C$ is equivalent to the application of the single operation $D$, and that $C$ followed by $A$ is equivalent to $F$. Such combinations of symmetry operations are denoted by the symbolic equations

$$CA = D, \quad AC = F$$

The results of the possible combinations of pairs of operations are summarized in the combination (or multiplication) table 2.1. The six operations form a closed set called a *group*. We shall use this group as a

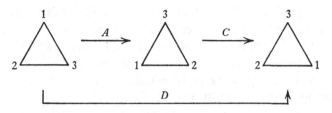

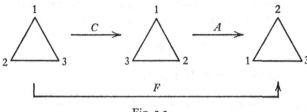

Fig. 2.2

convenient example in the following discussion, and refer to it as the group $G = \{E, A, B, C, D, F\}$.

TABLE 2.1

|   | E | A | B | C | D | F (applied first) |
|---|---|---|---|---|---|---|
| E | E | A | B | C | D | F |
| A | A | B | E | F | C | D |
| B | B | E | A | D | F | C |
| C | C | D | F | E | A | B |
| D | D | F | C | B | E | A |
| F | F | C | D | A | B | E |

**2.2.1** *Axioms of group theory*

A set of elements $\{E, P, Q, R, \dots\}$ forms a group if the following conditions are satisfied:

(i) The combination of any pair of elements of the group also belongs to the group

The law of combination depends on the nature of the elements; for example, addition or multiplication if the elements are numbers, matrix multiplication if they are matrices, or consecutive application of symmetry or other operations. The combination of two elements, $P$ and $Q$,

15

is called the product of $P$ and $Q$, and is written as $PQ$, with some convention about the ordering of the elements. The associative law of combination must hold for all the elements of the group, $P(QR) = (PQ)R = PQR$. The commutative law does not necessarily hold, and $PQ \neq QP$ in general; an example, for the group $G$, is $AC \neq CA$. If $PQ = QP$ for all elements of the group, the group is called an Abelian group.

(ii) One of the elements of the group, denoted by $E$, has the properties of a unit (or identity) element: for any element $P$,

$$PE = EP = P$$

(iii) Each element has an inverse which also belongs to the group: if $P$ belongs to the group then its inverse $P^{-1}$ is defined by

$$PP^{-1} = P^{-1}P = E$$

For the group $G$ the inverse elements are

$$E^{-1} = E, \quad A^{-1} = B, \quad B^{-1} = A, \quad C^{-1} = C, \quad D^{-1} = D, \quad F^{-1} = F$$

**2.2.2** *Some properties of groups*

The number of elements in a group is called the order $g$ of the group. The simplest group is the trivial group of order 1 which contains only the identity $E$. In addition to this trivial group, a group of order $g > 2$ can contain a number of other *subgroups* whose elements are subsets of the $g$ elements of the parent group. For example, the group

$$G = \{E, A, B, C, D, F\}$$

of order 6 contains the subgroups $\{E\}$, $\{E, C\}$, $\{E, D\}$, $\{E, F\}$ and $\{E, A, B\}$. The order of a subgroup is in general an integral factor of the order $g$ of the parent group (but there need not exist a subgroup for every integral factor).

The group $G$ contains three distinct sets of operations, $(E)$, $(A, B)$ and $(C, D, F)$, each of which consists of operations of the same type; each set forms a *class* It is often possible to pick out the different classes of symmetry operations from geometric considerations. For example, the operations $C$, $D$ and $F$ are rotations about geometrically equivalent axes and these axes, and hence the operations, can themselves be interchanged by a symmetry operation acting on the coordinate system to which the axes are tied. More generally, two elements $P$ and $Q$ belong to the same class if they are related by a *similarity transformation*

$$R^{-1}PR = Q$$

where $R$ is some element of the group and $R^{-1}$ is its inverse. For example, $F^{-1}CF = D$, and fig. 2.3 shows that this similarity transformation corresponds to an application of the operation $F$ to the coordinate system to which the axes $Oc$, $Od$, and $Of$ are tied, resulting in an interchange of the axes $Oc$ and $Od$ and, therefore, of the operations $C$ and $D$.

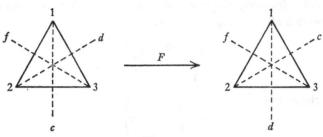

Fig. 2.3

If the group is Abelian, so that all the elements compute, then

$$R^{-1}PR = R^{-1}RP = P$$

and each element forms a class by itself. The number of elements in a class, its order, is in general an integral factor of the order of the group.

### 2.2.3  *Matrix representations*

A set of matrices that multiply in accordance with the multiplication table of a group is called a matrix representation $\Gamma$ of the group. Three representations of the group $\mathbf{G} = \{E, A, B, C, D, F\}$ are shown in table 2.2. $\Gamma_1$ and $\Gamma_2$ are one-dimensional representations, and are the only two sets of simple numbers which satisfy the multiplication table 2.1. $\Gamma_3$ is a two-dimensional representation. It is possible to construct an infinite number of representations of all dimensions for any group, but it can be shown that only a certain number of these can be regarded as distinct and independent. Thus, all the possible representations of the group $\mathbf{G}$ are either equivalent to or may be reduced to the three representations shown in table 2.2.

As an example of the construction of matrix representations, consider the vectors $\mathbf{r}_1, \mathbf{r}_2, \mathbf{r}_3$ joining the origin $O$ in fig. 2.1 to the points at the corners of the equilateral triangle. The application of a symmetry operation of the group $\mathbf{G}$ to these vectors results in a permutation of the vectors;

# Symmetry

TABLE 2.2

| | E | A | B | C | D | F |
|---|---|---|---|---|---|---|
| $\Gamma_1$ | 1 | 1 | 1 | 1 | 1 | 1 |
| $\Gamma_2$ | 1 | 1 | 1 | $-1$ | $-1$ | $-1$ |
| $\Gamma_3$ | $\begin{pmatrix} 1 & 0 \\ 0 & 1 \end{pmatrix}$ | $\begin{pmatrix} -\frac{1}{2} & -\sqrt{\frac{3}{4}} \\ \sqrt{\frac{3}{4}} & -\frac{1}{2} \end{pmatrix}$ | $\begin{pmatrix} -\frac{1}{2} & \sqrt{\frac{3}{4}} \\ -\sqrt{\frac{3}{4}} & -\frac{1}{2} \end{pmatrix}$ | $\begin{pmatrix} -1 & 0 \\ 0 & 1 \end{pmatrix}$ | $\begin{pmatrix} \frac{1}{2} & \sqrt{\frac{3}{4}} \\ \sqrt{\frac{3}{4}} & -\frac{1}{2} \end{pmatrix}$ | $\begin{pmatrix} \frac{1}{2} & -\sqrt{\frac{3}{4}} \\ -\sqrt{\frac{3}{4}} & -\frac{1}{2} \end{pmatrix}$ |

for example, $A\boldsymbol{r}_1 = \boldsymbol{r}_2$, $A\boldsymbol{r}_2 = \boldsymbol{r}_3$, and $A\boldsymbol{r}_3 = \boldsymbol{r}_1$. This permutation may be represented by the matrix equation

$$(\boldsymbol{r}_1 \boldsymbol{r}_2 \boldsymbol{r}_3) \begin{pmatrix} 0 & 0 & 1 \\ 1 & 0 & 0 \\ 0 & 1 & 0 \end{pmatrix} = (\boldsymbol{r}_2 \boldsymbol{r}_3 \boldsymbol{r}_1)$$

in which the matrix

$$A = \begin{pmatrix} 0 & 0 & 1 \\ 1 & 0 & 0 \\ 0 & 1 & 0 \end{pmatrix}$$

is a representation of the operation $A$. The six matrices obtained in this way form a three-dimensional representation of the group $\boldsymbol{G}$:

$$\left. \begin{aligned} E &= \begin{pmatrix} 1 & 0 & 0 \\ 0 & 1 & 0 \\ 0 & 0 & 1 \end{pmatrix} \quad A = \begin{pmatrix} 0 & 0 & 1 \\ 1 & 0 & 0 \\ 0 & 1 & 0 \end{pmatrix} \quad B = \begin{pmatrix} 0 & 1 & 0 \\ 0 & 0 & 1 \\ 1 & 0 & 0 \end{pmatrix} \\ C &= \begin{pmatrix} 1 & 0 & 0 \\ 0 & 0 & 1 \\ 0 & 1 & 0 \end{pmatrix} \quad D = \begin{pmatrix} 0 & 0 & 1 \\ 0 & 1 & 0 \\ 1 & 0 & 0 \end{pmatrix} \quad F = \begin{pmatrix} 0 & 1 & 0 \\ 1 & 0 & 0 \\ 0 & 0 & 1 \end{pmatrix} \end{aligned} \right\} \quad (2.1)$$

The vectors $\boldsymbol{r}_1, \boldsymbol{r}_2, \boldsymbol{r}_3$ are said to form a *basis* for this representation of the group.

From the representation $\Gamma = \{E, A, B, C, D, F\}$ defined by (2.1), it is possible to form a new representation $\Gamma' = \{E', A', B', C', D', F'\}$ in which each element $\boldsymbol{P}'$ is obtained from the corresponding element $\boldsymbol{P}$ of $\Gamma$ by the same similarity transformation

$$\boldsymbol{P}' = \boldsymbol{X}^{-1} \boldsymbol{P} \boldsymbol{X} \qquad (2.2)$$

where $\boldsymbol{X}$ is an arbitrary $(3 \times 3)$ matrix whose inverse is $\boldsymbol{X}^{-1}$. The new representation $\Gamma'$ satisfies the same group multiplication table as $\Gamma$. Thus, if $\boldsymbol{P}, \boldsymbol{Q}, \boldsymbol{R}$ are three elements of $\Gamma$ and $\boldsymbol{P}', \boldsymbol{Q}', \boldsymbol{R}'$ are the corresponding elements of $\Gamma'$, then $\boldsymbol{PQ} = \boldsymbol{R}$ implies $\boldsymbol{P}'\boldsymbol{Q}' = \boldsymbol{R}'$;

$$\boldsymbol{P}'\boldsymbol{Q}' = \boldsymbol{X}^{-1} \boldsymbol{P} \boldsymbol{X} \boldsymbol{X}^{-1} \boldsymbol{Q} \boldsymbol{X} = \boldsymbol{X}^{-1} \boldsymbol{P} \boldsymbol{Q} \boldsymbol{X} = \boldsymbol{X}^{-1} \boldsymbol{R} \boldsymbol{X} = \boldsymbol{R}'$$

Two representations that are related by a similarity transformation are called *equivalent* representations, and are considered as identical for group-theoretical purposes.

Consider now the representation $\Gamma'$ which is obtained from $\Gamma$ by means of the similarity transformation (2.2) with respect to the following matrix $X$:

$$X = \begin{pmatrix} \sqrt{\tfrac{1}{3}} & 0 & \sqrt{\tfrac{2}{3}} \\ \sqrt{\tfrac{1}{3}} & -\sqrt{\tfrac{1}{2}} & -\sqrt{\tfrac{1}{6}} \\ \sqrt{\tfrac{1}{3}} & \sqrt{\tfrac{1}{2}} & -\sqrt{\tfrac{1}{6}} \end{pmatrix} \quad X^{-1} = \begin{pmatrix} \sqrt{\tfrac{1}{3}} & -\sqrt{\tfrac{1}{3}} & \sqrt{\tfrac{1}{3}} \\ 0 & -\sqrt{\tfrac{1}{2}} & \sqrt{\tfrac{1}{2}} \\ \sqrt{\tfrac{2}{3}} & -\sqrt{\tfrac{1}{6}} & -\sqrt{\tfrac{1}{6}} \end{pmatrix}$$

The matrices of the new representation are

$$E' = \begin{pmatrix} 1 & 0 & 0 \\ 0 & 1 & 0 \\ 0 & 0 & 1 \end{pmatrix} \quad A' = \begin{pmatrix} 1 & 0 & 0 \\ 0 & -\tfrac{1}{2} & -\sqrt{\tfrac{3}{4}} \\ 0 & \sqrt{\tfrac{3}{4}} & -\tfrac{1}{2} \end{pmatrix} \quad B' = \begin{pmatrix} 1 & 0 & 0 \\ 0 & -\tfrac{1}{2} & \sqrt{\tfrac{3}{4}} \\ 0 & -\sqrt{\tfrac{3}{4}} & -\tfrac{1}{2} \end{pmatrix}$$

$$C' = \begin{pmatrix} 1 & 0 & 0 \\ 0 & -1 & 0 \\ 0 & 0 & 1 \end{pmatrix} \quad D' = \begin{pmatrix} 1 & 0 & 0 \\ 0 & \tfrac{1}{2} & \sqrt{\tfrac{3}{4}} \\ 0 & \sqrt{\tfrac{3}{4}} & -\tfrac{1}{2} \end{pmatrix} \quad F' = \begin{pmatrix} 1 & 0 & 0 \\ 0 & \tfrac{1}{2} & -\sqrt{\tfrac{3}{4}} \\ 0 & -\sqrt{\tfrac{3}{4}} & -\tfrac{1}{2} \end{pmatrix}$$

Each of these has the form

$$P' = \left( \begin{array}{c|c} P_1 & 0 \\ \hline 0 & P_3 \end{array} \right)$$

in which the $(2 \times 2)$ matrices $P_3$ form the two-dimensional representation $\Gamma_3$ in table 2.2, and the $(1 \times 1)$ matrices (numbers) $P_1$ form the one-dimensional representation $\Gamma_1$. The original representation $\Gamma$ is called a *reducible* representation, and $\Gamma'$ is its *reduced* form. The representations $\Gamma_1$ and $\Gamma_3$ cannot be reduced further by a similarity transformation and are called *irreducible* representations of the group. The reducible representation $\Gamma$ is called the direct sum of the irreducible representations $\Gamma_1$ and $\Gamma_3$; this is expressed symbolically as

$$\Gamma = \Gamma_1 + \Gamma_3$$

It can be shown that the three representations in table 2.2 are the only (non-equivalent) irreducible representations of the group $G$, so that every other representation is a direct sum of these three. In general, if

$$\Gamma = n_1\Gamma_1 + n_2\Gamma_2 + n_3\Gamma_3$$

the representation $\Gamma$ can be reduced by a similarity transformation to an equivalent representation $\Gamma'$ in which every matrix has the same block-diagonal form

$$P' = \begin{pmatrix} P_1 & & \mathbf{0} \\ & P_2 & \\ \mathbf{0} & & P_3 \\ & & & \ddots \end{pmatrix}$$

Non-zero elements appear only in the diagonal blocks, and the matrices $P_1, P_2, P_3$ of the irreducible representations $\Gamma_1, \Gamma_2, \Gamma_3$ occur $n_1, n_2, n_3$ times respectively.

When the order $g$ of a group is finite, the number and dimensions of the irreducible representations are governed by two general theorems:

(i) The number of non-equivalent irreducible representations is equal to the number of classes in the group.

(ii) The sum of the squares of the dimensions of the irreducible representations is equal to the order of the group:

$$\sum_{i=1}^{c} l_i^2 = l_1^2 + l_2^2 + \ldots + l_c^2 = g$$

where $c$ is the number of classes and $l_i$ is the dimension of the representation $\Gamma_i$. For the group $G$, $g = 6$ and $c = 3$, and there are therefore three irreducible representations whose dimensions are 1, 1 and 2 respectively.

Nearly all the applications of group theory in quantum chemistry rely on the properties of the irreducible representations.

### 2.2.4 *Group characters*

The character of an element $P$ in the representation $\Gamma$ of a group is the trace, or sum of the diagonal elements, of its representative matrix $P$ in this representation:

$$\chi(P) = \operatorname{tr} P = \sum_{n=1}^{l} P_{nn}$$

where $l$ is the dimension of $\Gamma$, and $P_{nn}$ $(n = 1, 2, \ldots, l)$ is a diagonal element of the matrix $P$. The character of the element $P$ in the irreducible representation $\Gamma_i$ is denoted by $\chi_i(P)$; for example, for the group $G$ (table 2.2),

$$\chi_1(C) = 1, \quad \chi_2(C) = -1, \quad \chi_3(C) = 0$$

Two matrices that are related by a similarity transformation have the same trace; from the rules of matrix multiplication,

$$\operatorname{tr} X^{-1} P X = \operatorname{tr} P X X^{-1} = \operatorname{tr} P$$

It follows that elements belonging to the same class have identical characters. Thus, for the group $G$,

$$\chi_i(C) = \chi_i(D) = \chi_i(F)$$

In addition, two equivalent representations, related by a similarity transformation, have the same system of characters, and this is the principal reason why equivalent representations are considered as the same for group-theoretical purposes.

TABLE 2.3  *Character table of the group* $G$

| $G$ | $E$ | $(A, B)$ | $(C, D, F)$ | |
|---|---|---|---|---|
| $\Gamma_1$ | 1 | 1 | 1 | |
| $\Gamma_2$ | 1 | 1 | $-1$ | $z$ |
| $\Gamma_3$ | 2 | $-1$ | 0 | $(x, y)$ |

The full set of characters of a group makes up the *character table* of the group. The character table of the group $G$ is table 2.3. The right-hand column of the table indicates the transformation properties of the co-ordinates of a point in the coordinate system defined by fig. 2.1. The $z$ coordinate is invariant with respect to the operations $A$ and $B$, but is antisymmetric with respect to $C, D$ and $F$; for example, $Az = z$, $Cz = -z$. The coordinate therefore belongs to (forms a basis for) the representation $\Gamma_2$. The coordinates $(x, y)$, on the other hand, belong to the representation $\Gamma_3$; they form a basis for the representation whose matrices are those shown in table 2.2. For example, in fig. 2.4, an anti-clockwise rotation through the angle $\theta$ about the $z$ axis moves a point from the position with coordinates $(x, y)$ to the new position $(x', y')$ where

$$x' = x \cos\theta - y \sin\theta$$
$$y' = x \sin\theta + y \cos\theta$$

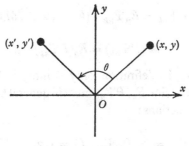

Fig. 2.4

## Symmetry

This transformation may be written in the matrix form

$$\begin{pmatrix} x' \\ y' \end{pmatrix} = \begin{pmatrix} \cos\theta & -\sin\theta \\ \sin\theta & \cos\theta \end{pmatrix} \begin{pmatrix} x \\ y \end{pmatrix}$$

and the $(2 \times 2)$ matrix is a representation of the rotation. For the symmetry operation $A$, $\theta = 2\pi/3$, $\cos\theta = -\frac{1}{2}$, $\sin\theta = \frac{1}{2}\sqrt{3}$, as in table 2.2

## 2.3 GROUP THEORY AND THE SCHRÖDINGER EQUATION

Consider the Schrödinger equation

$$\mathcal{H}\Psi = E\Psi \tag{2.3}$$

for the stationary states of an atom or molecule whose Hamiltonian is totally symmetric (invariant) with respect to a group of symmetry operations $\{P, Q, ...\}$. The application of a symmetry operation, $P$ say, to the left-hand side of (2.3) gives

$$P(\mathcal{H}\Psi) = \mathcal{H}(P\Psi)$$

since $P$ has no effect on $\mathcal{H}$. Application of $P$ to the right-hand side of (2.3) gives

$$P(E\Psi) = E(P\Psi)$$

since the energy is either a constant, for an atom, or a function of the internuclear distances in a (Born–Oppenheimer) molecule which is invariant with respect to interchanges of the coordinates of identical nuclei. Therefore, for any symmetric operation $P$,

$$\mathcal{H}(P\Psi) = E(P\Psi)$$

so that $\Psi$ and $P\Psi$ are eigenfunctions of $\mathcal{H}$ with the same eigenvalue $E$.

Let $E_n$ be an $M$-fold degenerate eigenvalue, and let the corresponding $M$ orthonormal eigenfunctions be $\Psi_{nk}$ $(k = 1, 2, ..., M)$;

$$\mathcal{H}\Psi_{nk} = E_n\Psi_{nk} \quad (k = 1, 2, ..., M)$$

Then

$$\mathcal{H}(P\Psi_{nk}) = E_n(P\Psi_{nk})$$

and because there are, by definition, only $M$ independent eigenfunctions belonging to the eigenvalue $E_n$, $P\Psi_{nk}$ must in general be a linear combination of these eigenfunctions:

$$P\Psi_{nk} = \sum_{l=1}^{M} P_{kl}\Psi_{nl} \quad (k = 1, 2, ..., M) \tag{2.4}$$

The $(M \times M)$ matrix $\boldsymbol{P}$ whose elements are the numbers $P_{kl}$ is a matrix representation of the operation $P$, and the collection of these matrices, one for each operation of the group, is an $M$-dimensional representation of the group. This representation can be shown to be *irreducible*.

This is one of the most important results of the application of group theory to quantum mechanics, and can be expressed in the form of a general theorem:

> *If a Hamiltonian is totally symmetric with respect to a group of symmetry operations then the eigenfunctions belonging to any eigenvalue form a basis for (belong to, or transform in accordance with) an irreducible representation of the group.*

If the eigenvalue is $M$-fold degenerate then the representation has dimension $M$. Conversely, the only possible degeneracies of eigenvalues (other than the rare 'accidental' degeneracies that are not related to symmetry) are given by the dimensions of the irreducible representations.

## 2.4 PROJECTION OPERATORS

Let the Hamiltonian of an atom or molecule be totally symmetric with respect to a group of $g$ symmetry operations $R_n$ with $c$ irreducible representations $\Gamma_i$. Let the dimension of the representation $\Gamma_i$ be $l_i$, and the character of the operation $R_n$ in $\Gamma_i$ be $\chi_i(R_n)$. With each irreducible representation there is associated a *projection operator*

$$\mathscr{P}_i = \frac{l_i}{g} \sum_{n=1}^{g} \chi_i(R_n)^* R_n \tag{2.5}$$

such that, given an arbitrary wave function $\Psi$ which is not an eigenfunction of the Hamiltonian and which does not have symmetry properties appropriate to an irreducible representation, the projected function

$$\Psi_i = \mathscr{P}_i \Psi = \frac{l_i}{g} \sum_{n=1}^{g} \chi_i(R_n)^* R_n \Psi \tag{2.6}$$

belongs to the representation $\Gamma_i$ and has the appropriate symmetry properties; it is a *symmetry-adapted* wave function. Projection operators provide a general and powerful method for the construction of wave functions with the correct symmetry properties for any state of an $N$-electron system.

An arbitrary (approximate) wave function $\Psi$ for the system may be expanded in terms of the complete set of eigenfunctions $\Psi_n$ of the Hamiltonian (§1.1),

$$\Psi = \sum_n C_n \Psi_n \qquad (2.7)$$

The effect of operating with $\mathscr{P}_i$ on $\Psi$ is to remove from this expansion all terms not belonging to the representation $\Gamma_i$. Thus, denoting the eigenfunctions belonging to $\Gamma_i$ by $\Psi_{n_i}$, (2.7) may be written as

$$\Psi = \sum_{n_1} C_{n_1} \Psi_{n_1} + \sum_{n_2} C_{n_2} \Psi_{n_2} + \dots + \sum_{n_c} C_{n_c} \Psi_{n_c} = \sum_{i=1}^{c} \Psi_i \qquad (2.8)$$

where

$$\Psi_i = \mathscr{P}_i \Psi = \sum_{n_i} C_{n_i} \Psi_{n_i}$$

is the component of $\Psi$ belonging to $\Gamma_i$. A second application of $\mathscr{P}_i$ has no further effect,

$$\mathscr{P}_i(\mathscr{P}_i \Psi) = \mathscr{P}_i \Psi \qquad (2.9)$$

whilst the application of $\mathscr{P}_j (j \neq i)$ annihilates $\Psi_i$,

$$\mathscr{P}_j(\mathscr{P}_i \Psi) = 0 \quad (j \neq i) \qquad (2.10)$$

The properties implied by (2.9) and (2.10),

$$\mathscr{P}_i^2 = \mathscr{P}_i, \quad \mathscr{P}_i \mathscr{P}_j = \mathscr{P}_j \mathscr{P}_i = 0 \quad (i \neq j) \qquad (2.11)$$

are characteristic properties of projection operators. In addition, (2.8) shows that the sum of the complete set of projection operators is equivalent to the identity operation,

$$\sum_{i=1}^{c} \Psi_i = \left( \sum_{i=1}^{c} \mathscr{P}_i \right) \Psi = \Psi \qquad (2.12)$$

## 2.5 ELECTRON PERMUTATION SYMMETRY. THE PAULI PRINCIPLE

The Hamiltonian (1.6) for a system of $N$ electrons is an operator which contains only terms involving the $3N$ spatial coordinates of the electrons, and it therefore has eigenfunctions which are functions of these $3N$ coordinates only. Because the electrons are identical particles, the Hamiltonian is invariant with respect to a permutation of the coordinates of the electrons, and its eigenfunctions must therefore have well-defined symmetry properties with respect to such permutations; they must transform in accordance with the irreducible representations of the permutation (or symmetric) group $\boldsymbol{P}_N$ whose group elements are the

*Electron permutation symmetry. The Pauli principle*

$N!$ permutations of $N$ objects (Wigner 1959, chapter 13). However, although it is possible to develop a spin-free form of quantum chemistry (Matsen 1964) it is convenient to include in a wave function information about the spins of the electrons. This is achieved most readily by the introduction of a fourth coordinate $\sigma$ for each electron which formally describes its spin motion, so that the wave function is now a function of $4N$ coordinates, four for each electron,

$$\Psi = \Psi(\boldsymbol{r}_1, \sigma_1, \boldsymbol{r}_2, \sigma_2, ..., \boldsymbol{r}_N, \sigma_N)$$

where $\boldsymbol{r}_i$ represents the three spatial coordinates of electron $i$ and $\sigma_i$ is its spin coordinate. This wave function must still belong to an irreducible representation of the permutation group, but it has been found that the properties of a system of electrons can be interpreted correctly only if the wave function satisfies the generalized Pauli principle; that is, if it belongs to the antisymmetric representation, with the property that it be antisymmetric with respect to the transposition of the (space and spin) coordinates of any pair of electrons:

$$\Psi(1, 2, ...i...j...N) = -\Psi(1, 2, ...j...i...N) \qquad (2.13)$$

for all $i \neq j$ and where, for example, the number 1 in parentheses indicates the dependence of the wave function on the four coordinates of electron 1. In the orbital approximation (chapter 3) the principle of antisymmetry reduces to the ordinary Pauli principle.

An arbitrary $N$-electron wave function which does not satisfy (2.13) can be made antisymmetric by means of the antisymmetrizing operator

$$\mathscr{A}_N = \frac{1}{N!} \sum_{i=1}^{N!} p_i P_i \qquad (2.14)$$

where $P_i$ is a permutation, and $p_i = +1$ or $-1$ according as $P_i$ corresponds to an even or odd number of particle transpositions. As shown in the following example, the numbers $p_i$ are the characters of the permutations in the antisymmetric representation of the permutation group, and $\mathscr{A}_N$ is the projection operator for this representation.

*Example. Three electrons.* A three-electron system belongs to the permutation group $\boldsymbol{P}_3$ whose elements are the six permutations of the three sets of electron coordinates. Any permutation can be expressed as a combination of transpositions $T_{ij}$ of the coordinates of two electrons at a time; in general,

$$T_{ij}\Psi(1, 2, ...i...j...N) = \Psi(1, 2, ...j...i...N)$$

*Symmetry*

and $T^2_{ij} = E$, the identity. The six operations of the group $P_3$ are

$$P_1 = E, \quad P_2 = T_{12}, \quad P_3 = T_{23},$$
$$P_4 = T_{13}, \quad P_5 = T_{12}T_{23}, \quad P_6 = T_{23}T_{12}$$

such that, for example,

$$P_2\Psi(1,2,3) = T_{12}\Psi(1,2,3) = \Psi(2,1,3)$$
$$P_5\Psi(1,2,3) = T_{12}T_{23}\Psi(1,2,3) = T_{12}\Psi(1,3,2) = \Psi(2,3,1)$$

These permutations satisfy the same combination table (table 2.1) as the operations of the group $G$ discussed in §2.2. The two groups, $P_3$ and $G$, are said to be *isomorphic*, and they have the same system of representations and characters (the groups are identical if the elements of $G$ are interpreted as permutations of the labels $(1,2,3)$ of the three points in fig. 2.1, with $A = P_5, B = P_6, C = P_3, D = P_4$, and $F = P_2$). The character table of $P_3$ is table 2.4.

TABLE 2.4 *Character table of the permutation group $P_3$*

| $P_3$ | $E$ | $(P_2, P_3, P_4)$ | $(P_5, P_6)$ |
|---|---|---|---|
| $\Gamma_1 = \Gamma_s$ | 1 | 1 | 1 |
| $\Gamma_2 = \Gamma_a$ | 1 | $-1$ | 1 |
| $\Gamma_3$ | 2 | 0 | $-1$ |

In common with all the permutation groups, $P_3$ has only two one-dimensional representations, the symmetric representation $\Gamma_s$ and the antisymmetric representation $\Gamma_a$. A wave function $\Psi_s$ belonging to $\Gamma_s$ is symmetric with respect to every transposition, $T_{ij}\Psi_s = \Psi_s$, and is therefore totally symmetric with respect to the permutations of the group. A wave function $\Psi_a$ belonging to $\Gamma_a$ is antisymmetric with respect to every transposition, $T_{ij}\Psi_a = -\Psi_a$. It is therefore symmetric with respect to every even permutation, equivalent to an even number of transpositions and with character $+1$ in $\Gamma_a$, and antisymmetric with respect to every odd permutation, with character $-1$ in $\Gamma_a$. The corresponding projection operator (2.5) for an $N$-electron system is the antisymmetrizer (2.14). For three electrons,

$$\mathscr{A}_3 = \frac{1}{3!}(P_1 - P_2 - P_3 - P_4 + P_5 + P_6)$$

26

and the antisymmetric projection of an arbitrary three-electron wave function $\Psi$ is

$$\mathscr{A}_3\Psi(1,2,3) = \frac{1}{3!}[\Psi(1,2,3) - \Psi(2,1,3) - \Psi(1,3,2) - \Psi(3,2,1)$$
$$+ \Psi(2,3,1) + \Psi(3,1,2)]$$

The two one-dimensional representations of the permutation group are the only representations of physical significance. Particles whose wave functions belong to the symmetric representation $\Gamma_s$ are called *bosons*; these are the particles, such as deuterons and $\alpha$-particles, that have integral or zero spin, and systems of these particles obey Bose–Einstein quantum statistics. Particles whose wave functions belong to the antisymmetric representation $\Gamma_a$ are called *fermions*; these are the particles, such as electrons, protons and neutrons, that have half-integral spin, and systems of these particles obey Fermi–Dirac statistics.

## 2.6 SPATIAL SYMMETRY

The Hamiltonian of an atom or molecule is invariant with respect to the spatial symmetry operations (rotations, reflections, etc.) that make up the symmetry point group of the system, and its eigenfunctions must therefore transform in accordance with the irreducible representations of the group. The construction of symmetry-adapted wave functions for non-linear molecules is readily achieved by the direct application of the projection operators (2.5), since the point groups of these molecules are of finite order. For an atom or a linear molecule however, the increased symmetry about one or more axes leads to point groups of infinite order, whose elements include the rotations through all possible angles about the axes, and a different, but related, approach is more appropriate.

### 2.6.1 *Molecular symmetry*

As an example of the spatial symmetry properties of non-linear molecules we consider the non-planar states of a symmetrical molecule $XY_3$, such as the ground state of $NH_3$. The molecule has a three-fold axis of rotation, $C_3$, and three equivalent $\sigma_v$ planes of symmetry, and it therefore belongs to the point group

$$C_{3v} = \{E, 2C_3, 3\sigma_v\}$$

where $2C_3$ represents the class of rotations $C_3$ and $C_3^2$ through angles $2\pi/3$ and $4\pi/3$ respectively about the axis, and $3\sigma_v$ represents the class of three reflections through the equivalent planes. Like the permutation group $P_3$, the point group $C_{3v}$ is isomorphic with the group $G$ discussed in §2.2, with the same system of representations (table 2.2) and characters (table 2.3).

TABLE 2.5  *Character table of the point group* $C_{3v}$

| $C_{3v}$ | $E$ | $2C_3$ | $3\sigma_v$ | |
|---|---|---|---|---|
| $A_1$ | 1 | 1 | 1 | $z$ |
| $A_2$ | 1 | 1 | $-1$ | |
| $E$ | 2 | $-1$ | 0 | $(x, y)$ |

The character table of $C_{3v}$ is table 2.5. The group has two one-dimensional representations, $A_1$ and $A_2$, and one two-dimensional representation, $E$, and the non-planar states of the molecule are therefore either non-degenerate or doubly degenerate. A wave function $\Psi^{A_1}$ belonging to the representation $A_1$ is totally symmetric with respect to the operations of the group:

$$C_3\Psi^{A_1} = C_3^2\Psi^{A_1} = \sigma_v\Psi^{A_1} = \sigma_v'\Psi^{A_1} = \sigma_v''\Psi^{A_1} = \Psi^{A_1}$$

A wave function $\Psi^{A_2}$ belonging to the representation $A_2$ is symmetric with respect to the class $2C_3$ but antisymmetric with respect to the class $3\sigma_v$:

$$C_3\Psi^{A_2} = C_3^2\Psi^{A_2} = \Psi^{A_2}$$

$$\sigma_v\Psi^{A_2} = \sigma_v'\Psi^{A_2} = \sigma_v''\Psi^{A_2} = -\Psi^{A_2}$$

Two degenerate wave functions $\Psi_1^E$ and $\Psi_2^E$, belonging to the representation $E$, are in general transformed into linear combinations of each other by the operations of the group; they form a basis for an irreducible representation ($E$) whose matrices are either identical to or equivalent to the matrices of the representation $\Gamma_3$ in table 2.2.

The projection operators for the three representations are

$$\left.\begin{aligned}
\mathscr{P}_{A_1} &= \tfrac{1}{6}(E + C_3 + C_3^2 + \sigma_v + \sigma_v' + \sigma_v'') \\
\mathscr{P}_{A_2} &= \tfrac{1}{6}(E + C_3 + C_3^2 - \sigma_v - \sigma_v' - \sigma_v'') \\
\mathscr{P}_E &= \tfrac{1}{3}(2E - C_3 - C_3^2)
\end{aligned}\right\} \quad (2.15)$$

and an arbitrary wave function for the system can be expressed as the sum of its projections, $\quad \Psi = \Psi^{A_1} + \Psi^{A_2} + \Psi^E$

where, for each representation $\Gamma$, $\Psi^\Gamma = \mathscr{P}_\Gamma\Psi$.

In the following example we consider the use of projection operators for the construction of symmetry-adapted molecular orbitals. We shall see in chapter 3, on the orbital approximation, that molecular orbitals are often calculated as eigenfunctions of a one-electron Hamiltonian which, like the full $N$-electron Hamiltonian of the molecule, is totally symmetric with respect to the symmetry operations of the molecular point group. The discussion in §2.3 of the transformation properties of wave functions then applies equally well to the molecular orbitals, and the relation between the dimensions of irreducible representations and the degeneracies of energies applies also to the orbital energies.

*Example. Symmetry orbitals for the ground state of* $NH_3$. In a simple LCAO–MO (§5.3) treatment of the bonding in the ground state of $NH_3$, the molecular orbitals are expressed as linear combinations of the $2s$ and $2p$ atomic orbitals on the nitrogen and the $1s$ orbital on each hydrogen. If the nitrogen nucleus is at the origin of a coordinate system whose $z$ axis is collinear with the $C_3$ axis of the molecule, the right-hand column of table 2.5 shows that the nitrogen $2p_x$ and $2p_y$ orbitals (which transform like $x$ and $y$ respectively) belong to the two-dimensional representation $E$, and the nitrogen $2p_z$ orbital belongs to the totally symmetric representation $A_1$, as does the nitrogen $2s$ orbital. On the other hand, the hydrogen orbitals do not individually have symmetry properties appropriate to the group $C_{3v}$†, but suitable symmetry-adapted combinations of them are readily constructed by means of the projection operators (2.15). Denoting the $1s$ orbitals on the hydrogens by $h_1, h_2, h_3$, we have, for example,

$$(C_3 + C_3^2) h_1 = h_2 + h_3$$

$$(\sigma_v + \sigma_v' + \sigma_v'') h_1 = h_1 + h_2 + h_3$$

and the application of the projection operators (2.15) to the orbitals gives the (unnormalized) symmetry functions

$$\phi_{A_1} = \mathscr{P}_{A_1} h_1 = \mathscr{P}_{A_1} h_2 = \mathscr{P}_{A_1} h_3 = (h_1 + h_2 + h_3)/3$$

$$\phi_{A_2} = \mathscr{P}_{A_2} h_1 = \mathscr{P}_{A_2} h_2 = \mathscr{P}_{A_2} h_3 = 0$$

$$\phi_E = \mathscr{P}_E h_1 = (2h_1 - h_2 - h_3)/3$$

$$\phi_E' = \mathscr{P}_E h_2 = (2h_2 - h_3 - h_1)/3$$

$$\phi_E'' = \mathscr{P}_E h_3 = (2h_3 - h_1 - h_2)/3$$

† The hydrogen orbitals form a basis for the three-dimensional representation of the group $C_{3v}$ whose matrices are identical to the matrices (2.1) of the reducible representation $\Gamma$ of the group $G$ discussed in §2.2.3.

Of the three functions belonging to the representation $E$, only two are linearly independent; for example,

$$\phi_E'' = -(\phi_E' + \phi_E)$$

so that three independent symmetry combinations of $h_1$, $h_2$ and $h_3$ are $\phi_{A_1}$, $\phi_E$ and $\phi_E'$.

In a conventional LCAO–MO calculation, the nitrogen $2s$ and $2p_z$ orbitals mix with $\phi_{A_1}$ to form three totally symmetric (delocalized) molecular orbitals belonging to $A_1$, whilst the $2p_x$ and $2p_y$ orbitals mix with $\phi_E$ and $\phi_E'$ to form two pairs of degenerate molecular orbitals belonging to $E$. The resulting orbital energy level diagram and ground state orbital configuration are shown schematically in fig. 2.5 (the $1a_1$ molecular orbital is essentially the nitrogen $1s$ atomic orbital).

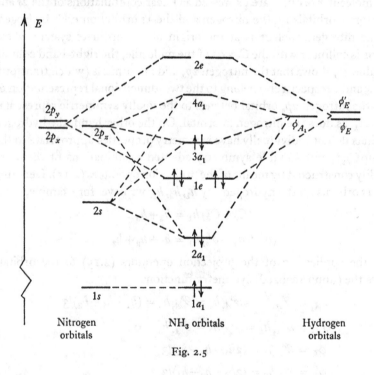

Fig. 2.5

### 2.6.2 *Angular momentum*

The cylindrical symmetry about the molecular axis of a linear molecule results in the quantization of the component of the total (spatial) angular momentum of the electrons along this axis (for a discussion of the relation between angular momentum and rotations about an axis, see McWeeny 1973, chapter 5). Let the axis lie along the $z$ direction. In classical mechanics the $z$ component of the angular momentum of a particle about a given origin is

$$L_z = xp_y - yp_x$$

where $p_x$ and $p_y$ are the $x$ and $y$ components of the linear momentum. The corresponding (Schrödinger) quantum-mechanical operator is obtained in the usual way by replacing each linear momentum variable $p$ by the differential operator $(\hbar/i)(\partial/\partial q)$, where $i = \sqrt{-1}$ and $q$ is the coordinate conjugate to momentum $p$; for example, $p_x$ is replaced by $(\hbar/i)(\partial/\partial x)$. The operator for the $z$ component of the angular momentum of a particle is therefore

$$\hbar\mathscr{L}_z = \frac{\hbar}{i}\left(x\frac{\partial}{\partial y} - y\frac{\partial}{\partial x}\right)$$

For $N$ particles, 
$$\mathscr{L}_z(1, 2, ..., N) = \sum_{k=1}^{N} \mathscr{L}_z(k)$$

where 
$$\mathscr{L}_z(k) = \frac{1}{i}\left(x_k\frac{\partial}{\partial y_k} - y_k\frac{\partial}{\partial x_k}\right)$$

In a linear molecule the quantization of the component of angular momentum along the molecular ($z$) axis means that the wave functions for the molecule are eigenfunctions of $\mathscr{L}_z$:

$$\mathscr{L}_z\Psi = M\Psi \quad (M = 0, \pm 1, \pm 2, ...)$$

The eigenvalue (quantum number) $M$ is the value, in (atomic) units of $\hbar$, of the component of angular momentum in the state $\Psi$ of the molecule. In an atom the spherical symmetry results in the quantization both of the square of the angular momentum of the electrons and of a component along a specified direction (conventionally chosen to be the $z$ direction):

$$\left.\begin{aligned}\mathscr{L}^2\Psi &= L(L+1)\Psi \quad (L = 0, 1, 2, ...) \\ \mathscr{L}_z\Psi &= M_L\Psi \quad (M_L = 0, \pm 1, \pm 2, ..., \pm L)\end{aligned}\right\} \quad (2.16)$$

where $\mathscr{L}^2 = \mathscr{L}_x^2 + \mathscr{L}_y^2 + \mathscr{L}_z^2$.

Projection operators for angular momentum may be constructed from a consideration of the eigenvalue equations (2.16). Consider an arbitrary wave function $\Psi$ which is not an eigenfunction of $\mathscr{L}^2$ but which, in accordance with (2.12), can always be expanded in the form

$$\Psi = \sum_{L'=0}^{\infty} \Psi_{L'}$$

where $\Psi_{L'}$ is that component of $\Psi$ which is an eigenfunction of $\mathscr{L}^2$ with eigenvalue $L'$:

$$[\mathscr{L}^2 - L'(L'+1)]\Psi_{L'} = 0$$

The operator $[\mathscr{L}^2 - L'(L'+1)]$ is called an annihilation operator, and its effect on $\Psi$ is to remove the component $\Psi_{L'}$ of $\Psi$ belonging to the eigenvalue $L'$. An operator which annihilates all the components except one, $\Psi_L$ say, is the product (Löwdin 1956)

$$\mathscr{P}_L = \prod_{L' \neq L} \left\{ \frac{\mathscr{L}^2 - L'(L'+1)}{L(L+1) - L'(L'+1)} \right\} \tag{2.17}$$

in which the terms in the denominator have been included to make $\mathscr{P}_L$ a projection operator:

$$\mathscr{P}_L \Psi = \sum_{L'} \mathscr{P}_L \Psi_{L'} = \Psi_L$$

*Example. The $p^2$ orbital configuration.* In an orbital treatment of the carbon atom the orbital configuration $(1s)^2(2s)^2(2p)^2$ gives rise to three sets of states whose symmetries, $^1$S, $^3$P and $^1$D, correspond to $L = 0, 1$ and $2$ respectively. The $s$ orbitals are spherically symmetric and are of no interest insofar as the (spatial) symmetry properties of the wave functions are concerned, and it is sufficient to consider only the symmetry properties of the $p^2$ configuration. We review first some general properties of the angular momentum operators and their eigensolutions.

An atomic orbital $\psi_{nlm}(\boldsymbol{r})$, which is an eigenfunction of the (one-electron) operators $\mathscr{L}^2$ and $\mathscr{L}_z$ with eigenvalues $l$ and $m$, can be expressed in spherical polar coordinates $(r, \theta, \phi)$ as the product of a radial function $R_{nl}(r)$ and a spherical harmonic $Y_{lm}(\theta, \phi)$, the latter determining the symmetry and angular momentum properties of the orbital:

$$\psi_{nlm}(\boldsymbol{r}) = R_{nl}(r)\, Y_{lm}(\theta, \phi)$$

and

$$\mathscr{L}^2 Y_{lm} = l(l+1)\, Y_{lm} \quad (l = 0, 1, 2, \ldots)$$

$$\mathscr{L}_z Y_{lm} = m Y_{lm} \quad (m = 0, \pm 1, \pm 2, \ldots, \pm l)$$

($\mathscr{L}^2$ and $\mathscr{L}_z$ operate only on the angular factor of the orbital). For the hydrogen atom $n$ is a quantum number which determines the energy, but in general it is simply a label which distinguishes different atomic orbitals with the same values of $l$ and $m$. A $p$ orbital has the form

$$p_m(\boldsymbol{r}) = R(r)\, Y_{1m}(\theta, \phi)$$

Because $\mathscr{L}^2 = \mathscr{L}_x^2 + \mathscr{L}_y^2 + \mathscr{L}_z^2$, it is necessary, for what follows, to know the result of operating on an atomic orbital with $\mathscr{L}_x$ and $\mathscr{L}_y$. It is in fact more convenient to work with the step operators

$$\mathscr{L}_+ = (\mathscr{L}_x + i\mathscr{L}_y), \quad \mathscr{L}_- = (\mathscr{L}_x - i\mathscr{L}_y)$$

in terms of which

$$\mathscr{L}^2 = \mathscr{L}_- \mathscr{L}_+ + \mathscr{L}_z^2 + \mathscr{L}_z$$

The step operators have the interesting property of changing the value of the quantum number $m$ (in general, of $M_L$) by $\pm 1$:

$$\left.\begin{array}{l} \mathscr{L}_+ Y_{lm} = [(l+m+1)(l-m)]^{\frac{1}{2}} Y_{l,\,m+1} \\ \mathscr{L}_- Y_{lm} = [(l+m)(l-m+1)]^{\frac{1}{2}} Y_{l,\,m-1} \end{array}\right\} \quad (2.18)$$

For a two-electron system,

$$\mathscr{L}^2(1,2) = \mathscr{L}_-(1,2)\,\mathscr{L}_+(1,2) + \mathscr{L}_z^2(1,2) + \mathscr{L}_z(1,2)$$

where $\mathscr{L}_\pm(1,2)$ and $\mathscr{L}_z(1,2)$ are sums of one-electron operators of the form

$$\mathscr{L}(1,2) = \mathscr{L}(1) + \mathscr{L}(2)$$

such that

$$\mathscr{L}(1,2)\,Y_{lm}(1)\,Y_{l'm'}(2) = [\mathscr{L}(1)\,Y_{lm}(1)]\,Y_{l'm'}(2) + Y_{lm}(1)\,[\mathscr{L}(2)\,Y_{l'm'}(2)] \quad (2.19)$$

In particular, if $\mathscr{L} = \mathscr{L}_z$, (2.19) shows that the product $Y_{lm}(1)\,Y_{l'm'}(2)$ is an eigenfunction of $\mathscr{L}_z(1,2)$ with eigenvalue $M_L = m + m'$.

TABLE 2.6

| $p_m p_{m'}$ | $p_1 p_1$ | $(p_1 p_0, p_0 p_1)$ | $(p_1 p_{-1}, p_{-1} p_1)$ | $p_0 p_0$ | $(p_0 p_{-1}, p_{-1} p_0)$ | $p_{-1} p_{-1}$ |
|---|---|---|---|---|---|---|
| $M_L = m + m'$ | 2 | 1 | 0 | 0 | $-1$ | $-2$ |
| $L$ | 2 | (1, 2) | (0, 1, 2) | (0, 2) | (1, 2) | 2 |

Returning to our example, the nine two-electron orbital products $p_m(1)\,p_{m'}(2)$, for $m, m' = 1, 0, -1$, are all eigenfunctions of $\mathscr{L}_z$ with eigenvalues $M_L = m + m'$, but they are not all eigenfunctions of $\mathscr{L}^2$. The third row of table 2.6 shows that only the functions $p_1 p_1$ and $p_{-1} p_{-1}$, having the maximum value of $|M_L| = 2$, are already eigenfunctions of $\mathscr{L}^2$, with eigenvalue $L = 2$. On the other hand, the function $p_0 p_0$, for example, has components belonging to the two eigenvalues $L = 0, 2$. An eigenfunction of $\mathscr{L}^2$ belonging to $L = 0$, say, may now be constructed from $p_0 p_0$ by annihilating the unwanted component belonging to $L = 2$. This is readily achieved by means of a projection operator $\mathscr{P}_0$, of the form (2.17), containing only the factor for $L' = 2$:

$$\mathscr{P}_0(1,2)\,p_0(1)\,p_0(2) = \left\{\frac{\mathscr{L}^2(1,2) - 2\cdot 3}{0 - 2\cdot 3}\right\} p_0(1)\,p_0(2) \qquad (2.20)$$

Since $\mathscr{L}_z(1,2)\,p_0(1)\,p_0(2) = 0$, we have

$$\mathscr{L}^2(1,2)\,p_0(1)\,p_0(2) = \mathscr{L}_-(1,2)\,\mathscr{L}_+(1,2)\,p_0(1)\,p_0(2)$$
$$= 2[2p_0(1)\,p_0(2) + p_1(1)\,p_{-1}(2) + p_{-1}(1)\,p_1(2)]$$

The resulting eigenfunction (2.20) belonging to $L = 0$ is (after normalization)

$$(1/\sqrt{3})[p_0(1)\,p_0(2) - p_1(1)\,p_{-1}(2) - p_{-1}(1)\,p_1(2)]$$
$$= (1/\sqrt{3})[p_x(1)\,p_x(2) + p_y(1)\,p_y(2) + p_z(1)\,p_z(2)]$$

the form on the right showing clearly that the function is spherically symmetric, as befits an S-state wave function. The corresponding (spatial) wave function for the lowest ¹S state of carbon is

$$(1/\sqrt{3})\,1s(1)\,1s(2)\,2s(3)\,2s(4)\,[2p_0(5)\,2p_0(6)$$
$$- 2p_1(5)\,2p_{-1}(6) - 2p_{-1}(5)\,2p_1(6)] \qquad (2.21)$$

## 2.7 SPIN SYMMETRY

The Hamiltonian (1.6) does not include spin-dependent terms and its eigenfunctions are therefore simultaneous eigenfunctions of the total spin angular momentum operators:

$$\mathscr{S}^2\Psi = S(S+1)\Psi$$
$$\mathscr{S}_z\Psi = M_s\Psi$$

For a single electron the values of the spin quantum numbers are $s = \frac{1}{2}$ and $m_s = \pm\frac{1}{2}$. The corresponding two spin states are conventionally described by the spin functions $\alpha(\sigma)$ for $m_s = +\frac{1}{2}$ and $\beta(\sigma)$ for $m_s = -\frac{1}{2}$, with the orthonormality properties

$$\int \alpha^*\alpha\,d\sigma = \int \beta^*\beta\,d\sigma = 1$$
$$\int \alpha^*\beta\,d\sigma = \int \beta^*\alpha\,d\sigma = 0$$

An $N$-electron spin function is the product of $N$ one-electron spin functions:

$$\eta_1(1)\,\eta_2(2)\dots\eta_N(N)$$

where $\eta_n(i) = \alpha(i)$ or $\beta(i)$ depending on the spin of electron $i$. It is in general possible to form $2^N$ such products, all of which are eigenfunctions of $\mathscr{S}_z$, the value of $M_S$ being the sum of the values of $m_s$ of the factors.

The construction of eigenfunctions of spin may be achieved by the same methods as are used for orbital angular momentum. Thus, corresponding to (2.17) we have the spin projection operators

$$\mathscr{P}_S = \prod_{S' \neq S} \left\{ \frac{\mathscr{S}^2 - S'(S'+1)}{S(S+1) - S'(S'+1)} \right\}$$

and the step operators are

$$\mathscr{S}_+ = (\mathscr{S}_x + i\mathscr{S}_y), \quad \mathscr{S}_- = (\mathscr{S}_x - i\mathscr{S}_y)$$

For the one-electron spin functions,

$$\mathscr{S}_z \alpha = +\tfrac{1}{2}\alpha, \quad \mathscr{S}_z \beta = -\tfrac{1}{2}\beta$$

and the equations corresponding to (2.18) are

$$\mathscr{S}_+ \alpha = 0, \quad \mathscr{S}_+ \beta = \alpha$$

$$\mathscr{S}_- \alpha = \beta, \quad \mathscr{S}_- \beta = 0$$

*Example. Three-electron spin functions.* Consider the set of eight three-electron spin functions shown in table 2.7. The functions $\alpha\alpha\alpha$ and $\beta\beta\beta$, having the maximum possible value of $|M_S| = \tfrac{3}{2}$, are already eigenfunctions of $\mathscr{S}^2$ with eigenvalue $S = \tfrac{3}{2}$. On the other hand, the functions $\alpha\alpha\beta$ and $\alpha\beta\beta$ have components which belong to both the eigenvalues $S = \tfrac{1}{2}, \tfrac{3}{2}$.

TABLE 2.7

| $\eta_1\eta_2\eta_3$ | $\alpha\alpha\alpha$ | $(\alpha\alpha\beta, \alpha\beta\alpha, \beta\alpha\alpha)$ | $(\alpha\beta\beta, \beta\alpha\beta, \beta\beta\alpha)$ | $\beta\beta\beta$ |
|---|---|---|---|---|
| $M_S = \Sigma m_s$ | $\tfrac{3}{2}$ | $\tfrac{1}{2}$ | $-\tfrac{1}{2}$ | $-\tfrac{3}{2}$ |
| $S$ | $\tfrac{3}{2}$ | $(\tfrac{3}{2}, \tfrac{1}{2})$ | $(\tfrac{3}{2}, \tfrac{1}{2})$ | $\tfrac{3}{2}$ |

Given $\alpha\alpha\alpha$, belonging to the quantum numbers $(S, M_S) = (\tfrac{3}{2}, \tfrac{3}{2})$, the step-down operator may be used to obtain the other eigenfunctions belonging to $S = \tfrac{3}{2}$. Thus

$$\mathscr{S}_-(\alpha\alpha\alpha) = (\mathscr{S}_-\alpha)\alpha\alpha + \alpha(\mathscr{S}_-\alpha)\alpha + \alpha\alpha(\mathscr{S}_-\alpha)$$

$$= \beta\alpha\alpha + \alpha\beta\alpha + \alpha\alpha\beta$$

has $(S, M_S) = (\tfrac{3}{2}, \tfrac{1}{2})$. A second application of $\mathscr{S}_-$ gives the function with

*Symmetry*

$(S, M_S) = (\frac{3}{2}, -\frac{1}{2})$, and a third application retrieves (a multiple of) $\beta\beta\beta$. The functions belonging to $(S, M_S) = (\frac{3}{2}, \pm\frac{1}{2})$ may alternatively be obtained from $\alpha\alpha\beta$ and $\alpha\beta\beta$ by annihilating the components belonging to $S = \frac{1}{2}$ by means of the projection operator

$$\mathscr{P}_{\frac{3}{2}} = \left\{ \frac{\mathscr{S}^2 - \frac{1}{2} \cdot \frac{3}{2}}{\frac{3}{2} \cdot \frac{5}{2} - \frac{1}{2} \cdot \frac{3}{2}} \right\}$$

*A special case.* The normalized singlet $(S = 0)$ two-electron spin function is

$$H_{0,0}(1, 2) = (1/\sqrt{2})[\alpha(1)\beta(2) - \beta(1)\alpha(2)] \qquad (2.22)$$

where, in general, $H_{S, M_S}$ denotes a spin eigenfunction with quantum numbers $S, M_S$. A product of $N$ of these spin eigenfunctions,

$$H_{0,0}(1, 2, ..., 2N) = H_{0,0}(1, 2) H_{0,0}(3, 4) ... H_{0,0}(2N-1, 2N) \qquad (2.23)$$

is again a spin eigenfunction with $S = 0$, and describes a $2N$-electron system in which all the spins are singlet-paired. If this function is now multiplied by an $M$-electron spin eigenfunction $H_{S, M_S}$ the resulting $(2N+M)$-electron function is again a spin eigenfunction with quantum numbers $S, M_S$:

$$H_{S, M_S}(1, 2, ..., 2N+M) = H_{0,0}(1, 2, ..., 2N)$$
$$\times H_{S, M_S}(2N+1, 2N+2, ..., 2N+M) \qquad (2.24)$$

This type of spin function is particularly important in the orbital approximation. As an example, we consider the lowest $^1S$ state of the carbon atom for which the spatial wave function (2.21) is

$$\Phi(1, 2, ..., 6) = (1/\sqrt{3}) 1s(1) 1s(2) 2s(3) 2s(4)$$
$$\times [2p_0(5) 2p_0(6) - 2p_1(5) 2p_{-1}(6) - 2p_{-1}(5) 2p_1(6)]$$

The (unnormalized) antisymmetrized orbital approximation for the $^1S$ state is then

$$\mathscr{A}_6 [\Phi(1, 2, ..., 6) H_{0,0}(1, 2, ..., 6)]$$

where $H_{0,0}$ is a singlet spin function of type (2.23). This wave function may be reduced to the form

$$(1/\sqrt{3}) \mathscr{A}_6 \{ 1s(1)\alpha(1) 1s(2)\beta(2) 2s(3)\alpha(3) 2s(4)\beta(4)$$
$$\times [2p_0(5) 2p_0(6) - 2p_1(5) 2p_{-1}(6) - 2p_{-1}(5) 2p_1(6)]\alpha(5)\beta(6) \} \qquad (2.25)$$

which describes an orbital configuration $(1s)^2 (2s)^2 (2p)^2$ in which the three pairs of electrons are singlet-coupled.

# 3 The orbital approximation

## 3.1 INTRODUCTION

The Schrödinger equation for an $N$-electron system is a partial differential equation in $3N$ variables, and is usually regarded as exactly soluble if the variables are separable; that is, if the equation in $3N$ variables can be transformed into $3N$ differential equations in one variable each. These can then be solved either analytically in terms of known functions, as in the case of the hydrogen atom, or numerically as in the case of the hydrogen molecule ion. For systems containing more than one electron however, the separation of variables is not possible because of the presence of the electron interaction terms $1/r_{ij}$ in the Hamiltonian (1.6),

$$\mathcal{H} = \sum_{i=1}^{N} \left( -\tfrac{1}{2}\nabla_i^2 - \sum_{a=1}^{\nu} \frac{Z_\alpha}{r_{i\alpha}} \right) + \sum_{i>j=1}^{N}\frac{1}{r_{ij}} + \sum_{a>\beta=1}^{\nu} \frac{Z_\alpha Z_\beta}{R_{\alpha\beta}}$$

In the absence of electron interaction the Hamiltonian (apart from the constant nuclear repulsion terms) reduces to a sum of one-electron operators,

$$\mathcal{H}_0 = \sum_{i=1}^{N} f(\boldsymbol{r}_i) \tag{3.1}$$

in which

$$f(\boldsymbol{r}_i) = -\tfrac{1}{2}\nabla_i^2 - \sum_{\alpha=1}^{\nu} \frac{Z_\alpha}{r_{i\alpha}} \tag{3.2}$$

is the Hamiltonian for a single electron in the presence of the stationary nuclei. The eigenfunctions of $\mathcal{H}_0$ are products of orbitals (one-electron wave functions)

$$\psi_1(\boldsymbol{r}_1)\,\psi_2(\boldsymbol{r}_2)\dots\Psi_N(\boldsymbol{r}_N) \tag{3.3}$$

and the orbitals are eigenfunctions of the one-electron operator $f$,

$$f(\boldsymbol{r})\,\psi_n(\boldsymbol{r}) = \epsilon_n \psi_n(\boldsymbol{r}) \tag{3.4}$$

The orbital product (3.3), after the inclusion of spin and antisymmetrization, is a simple orbital approximation for some state of the system. Quite generally, the orbital approximation is a model of an $N$-electron system in which the motion or 'state' of each electron is described by a *spin-orbital* which is a function of the coordinates of that electron only. Such a spin-orbital is usually assumed to have the form†

$$\phi_n(\boldsymbol{r}, \sigma) = \psi_n(\boldsymbol{r})\eta_n(\sigma) \tag{3.5}$$

† The most general type of spin-orbital has the form
$$\phi_n(\boldsymbol{r}, \sigma) = \psi_n^{(a)}(\boldsymbol{r})\,\alpha(\sigma) + \psi_n^{(\beta)}(\boldsymbol{r})\,\beta(\sigma)$$
This is not an eigenfunction of the spin operator $\mathcal{S}_z$, and an electron in such a spin-orbital does not have a well-defined spin. The general form is seldom used.

## The orbital approximation

in which the orbital $\psi_n$ describes the motion in space of the electron, and the spin function $\eta_n$ ($= \alpha$ or $\beta$) describes its spin state. The total wave function for an $N$-electron system is then an antisymmetrized product of $N$ spin-orbitals,

$$\Psi(1, 2, ..., N) = M' \mathscr{A}_N [\phi_1(1)\, \phi_2(2) ... \phi_N(N)] \qquad (3.6)$$

where $\phi_n(i) = \psi_n(i)\, \eta_n(i)$ is a function of the coordinates of electron $i$, and $M'$ is a number which can be chosen to normalize the wave function. This wave function is a linear combination of $N!$ spin-orbital products corresponding to the $N!$ ways of distributing the electrons amongst the $N$ spin-orbitals; for example, a three-electron wave function is (§2.5)

$$\Psi(1, 2, 3) = M[\phi_1(1)\, \phi_2(2)\, \phi_3(3) + \phi_1(2)\, \phi_2(3)\, \phi_3(1) + \phi_1(3)\, \phi_2(1)\, \phi_3(2)$$
$$- \phi_1(1)\, \phi_2(3)\, \phi_3(2) - \phi_1(2)\, \phi_2(1)\phi_3(3) - \phi_1(3)\, \phi_2(2)\, \phi_3(1)]$$

Such a wave function can be written in the form of a determinant, a *Slater determinant*:

$$\Psi(1, 2, ..., N) = M \begin{vmatrix} \phi_1(1) & \phi_1(2) & ... & \phi_1(N) \\ \phi_2(1) & \phi_2(2) & ... & \phi_2(N) \\ \multicolumn{4}{c}{\dotfill} \\ \phi_N(1) & \phi_N(2) & ... & \phi_N(N) \end{vmatrix}$$
$$= M \det [\phi_1(1)\, \phi_2(2) ... \phi_N(N)] \qquad (3.7)$$

which has all the usual properties of determinants. For example, the determinant is zero when two rows are identical, corresponding to the double occupancy of a spin-orbital. The principle of antisymmetry therefore reduces to the simple Pauli principle in the orbital approximation. In addition, the value of the determinant is unchanged when to any row is added a constant multiple of any other row; for example,

$$\det [(\phi_1(1) + \phi_2(1))\, \phi_2(2) ... \phi_N(N)] = \det [\phi_1(1)\, \phi_2(2) ... \phi_N(N)]$$

This shows that whereas a set of $N$ spin-orbitals uniquely defines the wave function $\Psi$ of (3.7), the converse is not true. Given a set of spin-orbitals, $\phi_n$, it is possible to construct an infinite number of alternative sets by the linear transformations

$$\phi'_m = \sum_{n=1}^{N} \phi_n C_{nm} \quad (m = 1, 2, ..., N) \qquad (3.8)$$

such that

$$\det\left[\phi_1'(1)\,\phi_2'(2)\dots\phi_N'(N)\right] = \det\left[\phi_1(1)\,\phi_2(2)\dots\phi_N(N)\right]$$

In particular, the set of spin-orbitals can always be chosen (in an infinite number of ways) to be orthogonal, as well as normalized:†

$$\int \phi_m^* \phi_n \, d\tau = \delta_{mn} \tag{3.9}$$

A Slater determinant (3.7) of $N$ orthonormal spin-orbitals is normalized when $M = (N!)^{-\frac{1}{2}}$, and will henceforth be denoted by

$$\Psi = |\phi_1, \phi_2, \dots, \phi_N|$$

The orthonormality of the spin-orbitals implies that the orbitals are themselves normalized, and that orbitals associated with the *same* spin factor are orthogonal. Thus, if

$$\phi_m(\boldsymbol{r}, \sigma) = \psi_m(\boldsymbol{r})\alpha(\sigma) \quad \text{and} \quad \phi_n(\boldsymbol{r}, \sigma) = \psi_n(\boldsymbol{r})\alpha(\sigma)$$

(3.9) becomes

$$\iint \phi_m^*(\boldsymbol{r}, \sigma)\,\phi_n(\boldsymbol{r}, \sigma)\,dv\,d\sigma = \int \psi_m^*(\boldsymbol{r})\,\psi_n(\boldsymbol{r})\,dv \int \alpha^*(\sigma)\alpha(\sigma)\,d\sigma$$

$$= \int \psi_m^*(\boldsymbol{r})\,\psi_n(\boldsymbol{r})\,dv = \delta_{mn}$$

On the other hand, orbitals associated with different spin factors are not necessarily orthogonal since the orthogonality of the spin-orbitals is then a consequence of the orthogonality of the spin functions,

$$\int \alpha^* \beta \, d\sigma = 0.$$

Thus, if $\phi_m = \psi_m \alpha$ and $\phi_n = \psi_n \beta$,

$$\int \phi_m^* \phi_n \, d\tau = \int \psi_m^* \psi_n \, dv \int \alpha^* \beta \, d\sigma = 0$$

Atomic and molecular wave functions are nearly always expressed in terms of orthogonal spin-orbitals, and we digress briefly to discuss one general method of constructing orthogonal functions.

---

† $\int \dots d\tau$ always implies integration over all relevant variables. In the present case, $d\tau = dv \cdot d\sigma$ and the integration is over the four coordinates of an electron. More generally, if the integrand is a function of the $4N$ space and spin coordinates of $N$ electrons, $d\tau = d\tau_1 d\tau_2 \dots d\tau_N$, where $d\tau_i = dv_i \cdot d\sigma_i$ is the space–spin volume element of the coordinates $(x_i, y_i, z_i, \sigma_i)$ of electron $i$.

## The orbital approximation

*Digression. The construction of orthogonal functions.* Consider a set of $M$ non-orthogonal functions $F_n$ $(n = 1, 2, ..., M)$, which may be spin-orbitals $\phi_n$, orbitals $\psi_n$, or $N$-electron wave functions $\Psi_n$. With these functions there is associated an 'overlap matrix' $S$, whose elements are the overlap integrals

$$S_{mn} = \int F_m^* F_n \, d\tau$$

A diagonal element $S_{nn}$ is unity if $F_n$ is normalized, and an off-diagonal element $S_{mn}$ is zero if $F_m$ and $F_n$ are orthogonal.

From the set $F_n$ it is possible to construct a new set of orthogonal functions $G_p$ by means of the linear transformation

$$G_p = \sum_{n=1}^{M} F_n C_{np} \quad (p = 1, 2, ..., M)$$

and this may be achieved in many ways. Perhaps the most popular method is the Schmidt orthogonalization method whereby the coefficients $C_{np}$ are chosen such that $C_{np} = 0$ if $n > p$:

$$G_1 = F_1 C_{11}$$
$$G_2 = F_1 C_{12} + F_2 C_{22}$$
$$G_3 = F_1 C_{13} + F_2 C_{23} + F_3 C_{33}$$

$$\cdots\cdots\cdots\cdots\cdots\cdots\cdots\cdots\cdots\cdots\cdots$$

$$G_M = F_1 C_{1M} + F_2 C_{2M} + ... + F_M C_{MM}$$

The procedure is to choose $G_1 = F_1$, and first to orthogonalize $F_2$ to $G_1$ to give $G_2$, then $F_3$ to $G_1$ and $G_2$ to give $G_3$, and so on. The general formula is

$$G_p = F_p - \sum_{n=1}^{p-1} G_n \frac{\int G_n^* F_p \, d\tau}{\int G_n^* G_n \, d\tau} \quad (p = 2, 3, ..., M)$$

The resulting orthogonal functions may then be normalized.

## 3.2 HARTREE–FOCK THEORY

Let the normalized Slater determinant

$$\Psi = |\phi_1, \phi_2, ..., \phi_N|$$

with orthonormal spin-orbitals $\phi_n(\boldsymbol{r}, \sigma) = \psi_n(\boldsymbol{r}) \eta_n(\sigma)$, be an approximate wave function for some state of an $N$-electron system whose Hamiltonian is (1.6). The corresponding energy of the system (in units of the Hartree energy $H_\infty$) is the expectation value of the Hamiltonian:

$$
\begin{aligned}
E &= \int \Psi^* \mathscr{H} \Psi \, d\tau \\
&= \sum_{n=1}^{N} f_n + \tfrac{1}{2} \sum_{m,\,n=1}^{N} (J_{mn} - K_{mn}) + \sum_{\alpha > \beta = 1}^{\nu} \frac{Z_\alpha Z_\beta}{R_{\alpha\beta}}
\end{aligned} \tag{3.10}
$$

The quantity $f_n$ represents the kinetic energy and nuclear-attraction energy of an electron in the orbital $\psi_n$:

$$
f_n = \int \phi_n^* f \phi_n \, d\tau = \int \psi_n^* f \psi_n \, dv \tag{3.11}
$$

where $f$ is the Hamiltonian (3.2) for the motion of a single electron in the presence of the stationary nuclei (the second integral in (3.11) has been obtained from the first by integration over the spin). $J_{mn}$ is called a *Coulomb integral* and represents the Coulomb energy of interaction of two charge distributions with densities $|\psi_m|^2$ and $|\psi_n|^2$; that is, of two electrons in the orbitals $\psi_m$ and $\psi_n$:

$$
\begin{aligned}
J_{mn} &= \iint \phi_m^*(1) \, \phi_n^*(2) \frac{1}{r_{12}} \phi_m(1) \, \phi_n(2) \, d\tau_1 d\tau_2 \\
&= \iint \psi_m^*(1) \, \psi_m(1) \frac{1}{r_{12}} \psi_n^*(2) \, \psi_n(2) \, dv_1 dv_2
\end{aligned} \tag{3.12}
$$

$K_{mn}$ is an *exchange integral* and represents the interaction of two electrons in the 'exchange' charge distribution $|\psi_m^* \psi_n|$:

$$
\begin{aligned}
K_{mn} &= \iint \phi_m^*(1) \, \phi_n^*(2) \frac{1}{r_{12}} \phi_n(1) \, \phi_m(2) \, d\tau_1 d\tau_2 \\
&= \iint \psi_m^*(1) \, \psi_n(1) \frac{1}{r_{12}} \psi_n^*(2) \, \psi_m(2) \, dv_1 dv_2 \cdot \left| \int \eta_m^* \eta_n \, d\sigma \right|^2
\end{aligned} \tag{3.13}
$$

and $K_{nn} = J_{nn}$. The exchange integral $K_{mn}$ is non-zero only if the two spin-orbitals have the same spin factor; that is, if $\eta_m = \eta_n$. Because the exchange integrals are non-negative (as are the Coulomb integrals), we see that the exchange interactions of electrons with like spins lead to a lowering of the energy and a stabilization of the state. This forms the basis, all other factors being equal, of Hund's rule of maximum spin multiplicity.

41

## The orbital approximation

*Example. The ground state of lithium.* An orbital approximation for the $^2S$ ground state of the lithium atom is the Slater determinant

$$\Psi = |\phi_1, \phi_2, \phi_3|$$

in which $\phi_1 = 1s \cdot \alpha$, $\phi_2 = 1s \cdot \beta$, and $\phi_3 = 2s \cdot \alpha$. This wave function describes a $(1s)^2 (2s)$ orbital configuration, and the corresponding energy is (since $K_{nn} = J_{nn}$)

$$E = f_1 + f_2 + f_3 + J_{12} + J_{13} + J_{23} - K_{12} - K_{13} - K_{23}$$

In terms of the orbitals, we have $f_1 = f_2 = f_{1s}$ and $f_3 = f_{2s}$, where, for example,

$$f_{1s} = \int 1s(r) f(r)\, 1s(r)\, dv$$

is the kinetic energy plus nuclear-attraction energy of an electron in the $1s$ orbital. Similarly, $J_{12} = J_{1s,1s}, J_{13} = J_{23} = J_{1s,2s}$, where, for example,

$$J_{1s,2s} = \int \int \frac{1s(1)^2\, 2s(2)^2}{r_{12}}\, dv_1\, dv_2$$

is the Coulomb energy of interaction between an electron in the $1s$ orbital and the electron in the $2s$ orbital. Of the exchange terms, only $K_{13} = K_{1s,2s}$ is non-zero because of the different spin factors of the spin-orbitals in $K_{12}$ and $K_{23}$;

$$K_{1s,2s} = \int \int \frac{1s(1)\, 2s(1)\, 1s(2)\, 2s(2)}{r_{12}}\, dv_1\, dv_2$$

is the exchange energy of interaction between the electron in the $2s$ orbital and an electron with the same spin in the $1s$ orbital. The energy of the atom is then

$$E = 2f_{1s} + f_{2s} + J_{1s,1s} + 2J_{1s,2s} - K_{1s,2s}$$

We have already seen that in the absence of electron interaction the orbitals are eigenfunctions of the operator $f$ which contains, in addition to the kinetic energy operator, only the potential energy of attraction between an electron and the nuclei. The resulting Slater determinant (3.7), which is an eigenfunction of $\mathscr{H}_0$, (3.1), can be regarded as an approximate wave function for some state of the system whose Hamiltonian is (1.6). In general, however, this wave function must be expected to give a rather poor description of the state since it fails to recognize in any way the interactions between the electrons. An improved description, within the orbital approximation, is obtained by keeping the form (3.7) of the total wave function, but using the variation principle to calculate

42

the orbitals. Thus, given a Slater determinant in which the orbitals contain a number of arbitrary parameters, 'best' orbitals of the given form may be calculated by minimizing the energy (3.10) with respect to the possible variations of the orbitals; that is, with respect to the parameters. As discussed in §1.3, increasing the number of variational parameters in the orbitals results in improved accuracy of the energy. As the number of parameters is increased indefinitely, both the energy and the wave function converge to well-defined limiting values which are the most accurate possible within the orbital approximation.

*Example. The ground state of helium.* An orbital approximation for the $^1S$ ground state of the helium atom is the Slater determinant

$$\Psi = |\phi_1, \phi_2|$$

in which $\phi_1 = 1s \cdot \alpha$, $\phi_2 = 1s \cdot \beta$. This wave function describes a $(1s)^2$ orbital configuration, and the corresponding energy is

$$E = 2f_{1s} + J_{1s,1s}$$

where $f_{1s}$ is the kinetic energy plus nuclear-attraction energy of an electron in the $1s$ orbital, and $J_{1s,1s}$ is the Coulomb term for the interaction of the two electrons. A suitable form of the orbital is

$$1s(r) = \sum_{i=1}^{N} C_i \chi_{1s}(r, \zeta_i)$$

where
$$\chi_{1s}(r, \zeta_i) = (\zeta_i^3/\pi)^{\frac{1}{2}} e^{-\zeta_i r}$$

is a normalized $1s$ orbital for a hydrogen-like atom with nuclear charge $\zeta_i e$. The corresponding energy is a function of the parameters $C_i$ and $\zeta_i$, and the best orbital of this form is obtained by minimizing the energy with respect to the parameters. The results for several values of $N$ are shown in table 3.1. Of the two one-term orbitals ($N = 1$), the first, with $\zeta_1 = 2$, is a solution of the Schrödinger equation for the one-electron atom He$^+$, whereas the second, with $\zeta_1 = 2 - \frac{5}{16}$, allows for electron interaction in the neutral atom by the inclusion of a screening factor. The energy obtained with the five-term orbital cannot be improved (to the given number of figures) by increasing $N$, and is the limiting value of the orbital approximation $(1s)^2$. The exact (non-relativistic) energy is $-2.903\,72H_\infty$, and the remaining error of $0.042H_\infty$ ($105\,\text{kJ mol}^{-1}$) is due to the deficiencies of the model. Methods for going beyond the orbital approximation are discussed in chapter 4.

*The orbital approximation*

TABLE 3.1   *The ground state of helium*

| $N$ | $-E/H_\infty$ | $\zeta_1$ | $\zeta_2$ | $\zeta_3$ | $\zeta_4$ | $\zeta_5$ |
|---|---|---|---|---|---|---|
| 1 | 2.75 | 2 | | | | |
| 1 | 2.84765 | 1.69 | | | | |
| 2 | 2.86167† | 1.45 | 2.86 | | | |
| 5 | 2.86168† | 1.43 | 2.44 | 4.10 | 6.48 | 0.80 |

† Clementi (1965 b)

It must be pointed out however that requiring the electrons to occupy the same orbital is a constraint on the wave function, and a lower, more accurate, energy can be obtained by allowing the electrons to occupy different orbitals. In addition, for many systems, the assumption of symmetry orbitals (in this case an $s$ orbital) is also a constraint.

The limiting form of the orbital approximation is obtained in general by means of the variation principle when the orbitals of (3.7) are allowed to vary without any constraints on their form and complexity, other than the requirement that they satisfy the necessary boundary conditions. Minimization of the energy (3.10) with respect to all possible variations of the orbitals, subject to the condition that the spin-orbitals be orthonormal (which we have seen is no constraint on the wave function), then leads quite generally to a one-electron eigenvalue equation of the form

$$h^{\mathrm{HF}}\phi_n = \epsilon_n \phi_n \qquad (3.14)$$

and the calculation of the orbitals is therefore reduced to the determination of the eigenfunctions of the one-electron *Hartree–Fock* Hamiltonian $h^{\mathrm{HF}}$. The corresponding eigenvalues

$$\epsilon_n = \int \phi_n^* h^{\mathrm{HF}} \phi_n \, \mathrm{d}\tau \qquad (3.15)$$

are interpreted as orbital energies. The resulting total wave function, $\Psi = |\phi_1, \phi_2, ..., \phi_N|$ is then an eigenfunction of the $N$-electron Hartree–Fock Hamiltonian

$$\mathscr{H}^{\mathrm{HF}} = \sum_{i=1}^{N} h^{\mathrm{HF}}(i)$$

and the corresponding eigenvalue is the sum of the orbital energies, $\sum_{n=1}^{N} \epsilon_n$. The operator $h^{\mathrm{HF}}$ differs from the 'bare-nucleus' operator $f$, (3.2),

44

in that it contains terms which describe in an average way the interaction between one electron and all the other electrons in the system:

$$h^{\mathrm{HF}} = f + \mathscr{J} - \mathscr{K} \tag{3.16}$$

where $\mathscr{J}$ is the *Coulomb operator*

$$\mathscr{J}(\mathrm{1}) = \sum_{n=1}^{N} \int \frac{\phi_n^*(2)\,\phi_n(2)}{r_{12}}\,\mathrm{d}\tau_2 \tag{3.17}$$

and the *exchange operator* $\mathscr{K}$ is defined by its operation on a spin-orbital,

$$\mathscr{K}(\mathrm{1})\,\phi_m(\mathrm{1}) = \sum_{n=1}^{N} \phi_n(\mathrm{1}) \int \frac{\phi_n^*(2)\,\phi_m(2)}{r_{12}}\,\mathrm{d}\tau_2 \tag{3.18}$$

### 3.2.1 *The self-consistent field method*

The expressions (3.17) and (3.18) show that the equation (3.14) is only formally an eigenvalue equation, since $h^{\mathrm{HF}}$ is a function of the (unknown) occupied spin-orbitals. Equation (3.14) represents a set of $N$ simultaneous non-linear equations, the *Hartree–Fock equations*, whose solution gives both the $N$ occupied spin-orbitals *and* the Hartree–Fock operator.

The most commonly used method of solving the Hartree–Fock equations is the self-consistent field (SCF) method, which can be summarized as

$$\ldots \phi_n^{(i)} \to h^{(i)} \to \phi_n^{(i+1)} \to h^{(i+1)} \ldots$$

An initial Hamiltonian, $h^{(0)}$ say, is calculated from some suitably chosen initial set of spin-orbitals $\phi_n^{(0)}$. The eigenfunctions $\phi_n^{(1)}$ of $h^{(0)}$ are then used to construct a new operator $h^{(1)}$ whose eigenfunctions are $\phi_n^{(2)}$. This iterative cycle is continued until the solutions are self-consistent; that is, until the eigenfunctions $\phi_n^{(i)}$ and eigenvalues $\epsilon_n^{(i)}$ are the same, to the required accuracy, as $\phi_n^{(i-1)}$ and $\epsilon_n^{(i-1)}$.

The Hartree–Fock operator $h^{\mathrm{HF}}$ obtained in this way has, in addition to the $N$ occupied spin-orbitals, an infinite number of other eigenfunctions, called *virtual, unoccupied* or *excited* spin-orbitals. The occupied and virtual spin-orbitals together form a complete set of orthonormal one-electron wave functions. The methods that have been developed for the practical (approximate) solution of the Hartree–Fock equations are discussed in chapter 5.

### 3.2.2 *Constrained and extended Hartree–Fock models*

Although an exact solution of the Hartree–Fock equations (3.14) would give the best possible single-determinant approximation to the ground-state wave function, such a completely general calculation is seldom performed in practice. Assumptions are usually made which simplify both

# The orbital approximation

the calculation and the interpretation of the orbitals and, because these assumptions lead to constraints on the form of the orbitals, the resulting wave function is not the most accurate possible within the orbital approximation.

Given that the spin-orbitals have the form (3.5),†

$$\phi_n(\mathbf{r}, \sigma) = \psi_n(\mathbf{r})\eta_n(\sigma)$$

one of the two most frequently made assumptions is that the spin-orbitals occur in pairs,

$$\psi_n(\mathbf{r})\alpha(\sigma) \quad \text{and} \quad \psi_n(\mathbf{r})\beta(\sigma)$$

with a common orbital but with different spin factors. This pairing of spin-orbitals, which is associated with the concept of the double occupancy of orbitals, is not generally exhibited by the spin-orbitals obtained by an *unconstrained* solution of the Hartree–Fock equations (3.14).

The Hartree–Fock operator (3.16) is a spin-dependent operator whose effect on a spin-orbital $\phi_n$ depends on the spin factor $\eta_n$ of $\phi_n$. Consider first the Coulomb operator (3.17). Separation of the space and spin factors gives

$$\mathscr{J}(1) = \sum_{n=1}^{N} \int \frac{\psi_n^*(2)\psi_n(2)}{r_{12}} \, dv_2 \int \eta_n^*(2)\eta_n(2) \, d\sigma_2$$

$$= \sum_{n=1}^{N} \int \frac{\psi_n^*(2)\psi_n(2)}{r_{12}} \, dv_2 \tag{3.19}$$

so that $\mathscr{J}$ is spin-independent. The exchange operator $\mathscr{K}$, on the other hand, is spin-dependent. Consider, for example, a three-electron system whose spin-orbitals are $\phi_1 = \psi_1\alpha$, $\phi_2 = \psi_2\beta$ and $\phi_3 = \psi_3\alpha$. The corresponding exchange operator is given by, (3.18),

$$\mathscr{K}(1)\phi_m(1) = \phi_1(1) \int \frac{\phi_1^*(2)\phi_m(2)}{r_{12}} \, d\tau_2 + \phi_2(1) \int \frac{\phi_2^*(2)\phi_m(2)}{r_{12}} \, d\tau_2$$

$$+ \phi_3(1) \int \frac{\phi_3^*(2)\phi_m(2)}{r_{12}} \, d\tau_2$$

or, separating the space and spin factors.

$$\mathscr{K}(1)\{\psi_m(1)\eta_m(1)\} = \psi_1(1)\alpha(1) \int \frac{\psi_1^*(2)\psi_m(2)}{r_{12}} \, dv_2 \int \alpha^*(2)\eta_m(2) \, d\sigma_2$$

$$+ \psi_2(1)\beta(1) \int \frac{\psi_2^*(2)\psi_m(2)}{r_{12}} \, dv_2 \int \beta^*(2)\eta_m(2) \, d\sigma_2$$

$$+ \psi_3(1)\alpha(1) \int \frac{\psi_3^*(2)\psi_m(2)}{r_{12}} \, dv_2 \int \alpha^*(2)\eta_m(2) \, d\sigma_2$$

$$\tag{3.20}$$

† This is already an assumption; see the footnote to p. 37.

If $\phi_m$ has the spin factor $\eta_m = \alpha$ ($\phi_m = \psi_m^{(\alpha)}\alpha$, say) then, because of the orthogonality of the spin functions, (3.20) reduces to

$$\mathscr{K}(1)\{\psi_m^{(\alpha)}(1)\alpha(1)\} = \left\{\psi_1(1)\int \frac{\psi_1^*(2)\,\psi_m^{(\alpha)}(2)}{r_{12}}\,dv_2\right.$$

$$\left. + \psi_3(1)\int \frac{\psi_3^*(2)\,\psi_m^{(\alpha)}(2)}{r_{12}}\,dv_2\right\}\alpha(1) = \{\mathscr{K}_\alpha(1)\psi_m^{(\alpha)}(1)\}\alpha(1)$$

where $\mathscr{K}_\alpha$ is a spin-independent exchange operator for the orbitals associated with $\alpha$ spin. On the other hand, if $\phi_m = \psi_m^{(\beta)}\beta$,

$$\mathscr{K}(1)\{\psi_m^{(\beta)}(1)\beta(1)\} = \left\{\psi_2(1)\int \frac{\psi_2^*(2)\,\psi_m^{(\beta)}(2)}{r_{12}}\,dv_2\right\}\beta(1)$$

$$= \{\mathscr{K}_\beta(1)\psi_m^{(\beta)}(1)\}\beta(1)$$

and $\mathscr{K}_\beta \neq \mathscr{K}_\alpha$. More generally, let $\psi_n^{(\alpha)}$ ($n = 1, 2, ..., N_\alpha$) be the $N_\alpha$ occupied orbitals that are associated with $\alpha$ spin, and let $\psi_n^{(\beta)}$ ($n = N_\alpha + 1$, $N_\alpha + 2, ..., N$) be the $(N - N_\alpha)$ occupied orbitals that are associated with $\beta$ spin. Then, if $\phi_m = \psi_m^{(\alpha)}\alpha$, (3.18) can be replaced by the spin-independent expression

$$\mathscr{K}_\alpha(1)\psi_m^{(\alpha)}(1) = \sum_{n=1}^{N_\alpha} \psi_n^{(\alpha)}(1)\int \frac{\psi_n^{(\alpha)*}(2)\,\psi_m^{(\alpha)}(2)}{r_{12}}\,dv_2 \qquad (3.21)$$

whereas if $\phi_m = \psi_m^{(\beta)}\beta$,

$$\mathscr{K}_\beta(1)\psi_m^{(\beta)}(1) = \sum_{n=N_\alpha+1}^{N} \psi_n^{(\beta)}(1)\int \frac{\psi_n^{(\beta)*}(2)\,\psi_m^{(\beta)}(2)}{r_{12}}\,dv_2 \qquad (3.22)$$

The two sets of orbitals therefore satisfy *different* eigenvalue equations:

$$\left.\begin{array}{l} h_\alpha^{\mathrm{HF}}\psi_n^{(\alpha)} = (f + \mathscr{J} - \mathscr{K}_\alpha)\psi_n^{(\alpha)} = \epsilon_n^{(\alpha)}\psi_n^{(\alpha)} \\[4pt] h_\beta^{\mathrm{HF}}\psi_n^{(\beta)} = (f + \mathscr{J} - \mathscr{K}_\beta)\psi_n^{(\beta)} = \epsilon_n^{(\beta)}\psi_n^{(\beta)} \end{array}\right\} \qquad (3.23)$$

As a consequence, no orbital $\psi_n^{(\alpha)}$ is in general the same as any orbital $\psi_n^{(\beta)}$, and the concept of the double occupancy of orbitals, so useful in chemistry, is not obtained naturally from the general (unconstrained) Hartree–Fock model, but must be introduced as an assumption about and a constraint on the orbitals.

The second most commonly made assumption is that the orbitals are symmetry orbitals, belonging to the various irreducible representations of the symmetry point group of the system. In general, this requires that both $h_\alpha^{\mathrm{HF}}$ and $h_\beta^{\mathrm{HF}}$ be totally symmetric with respect to the same group of symmetry operations as the full Hamiltonian $\mathscr{H}$. In many cases of interest in chemistry, particularly when the orbitals are allowed to be doubly occupied, choosing the initial set of orbitals in the SCF procedure to be symmetry orbitals is sufficient to ensure that the Hartree–Fock Hamiltonian be totally symmetric. In other cases it may be necessary, or convenient for computational and interpretative reasons, to use a Hamiltonian which is a totally symmetric average for a number of related states (see for example, McWeeny and Sutcliffe 1969, §5.4).

Even with the assumption of the double-occupancy and symmetry constraints on the orbitals, however, a single Slater determinant does not generally satisfy the overall space and spin symmetry requirements discussed in chapter 2. One way of overcoming this deficiency is first to calculate the orbitals of the single-determinant approximation, and then to use the orbitals to construct a new wave function (by means of projection operators, for example) with the correct symmetry properties of the state in question. Such a wave function will be a linear combination of a small number of Slater determinants. A second way is to construct the symmetry-adapted wave function *prior* to the calculation of the orbitals. We shall return to these extensions of the Hartree–Fock model in §3.3.2 and §3.6.

### 3.3 THE RESTRICTED HARTREE–FOCK MODEL

The restricted Hartree–Fock (RHF) model is the simplest and by far the most widely used form of Hartree–Fock theory, and it forms the basis of what is commonly known as 'molecular-orbital theory'. The model involves the assumption of both the double-occupancy and symmetry constraints.

Let the Hamiltonian of an atom or molecule be totally symmetric with respect to the spatial symmetry operations of a point group whose irreducible representations $\Gamma_i$ have dimensions $l_i$.† The symmetry constraint then implies that the orbitals occur in sets of $l_i$ spatially degenerate symmetry orbitals belonging to the representations $\Gamma_i$. When every member of such a degenerate set is singly occupied and is associated with the same spin, $\alpha$ or $\beta$, then the set makes a totally symmetric contribution to the corresponding Hartree–Fock operator, $h_\alpha^{HF}$ or $h_\beta^{HF}$ of (3.23). When a given set of spatially degenerate orbitals is associated with *both* the $\alpha$-spin and $\beta$-spin factors, the corresponding $2l_i$ degenerate spin-orbitals are said to form a *shell*. When every orbital of the set is doubly occupied, the shell is said to be *closed*.

---

† For an atom the representations are associated with the angular momentum quantum number $L$, and the degeneracies are $(2L + 1)$, corresponding to the possible values of $M_L$.

### 3.3.1 *Closed-shell states*

A closed-shell state of a system of $N$ electrons is a totally symmetric singlet state which can be represented by a single Slater determinant made up of closed shells only; that is, of $N/2$ orbitals with every member of each spatially degenerate set doubly occupied:

$$
\begin{aligned}
\Psi'(1, 2, \ldots, N) &= (N!)^{-\frac{1}{2}} \det[\psi_1(1)\alpha(1)\psi_1(2)\beta(2)\psi_2(3)\alpha(3)\psi_2(4)\beta(4)\cdots \\
&\quad \times \psi_{N/2}(N-1)\alpha(N-1)\psi_{N/2}(N)\beta(N)] \\
&= |\psi_1\alpha, \psi_1\beta, \psi_2\alpha, \psi_2\beta, \ldots, \psi_{N/2}\alpha, \psi_{N/2}\beta|
\end{aligned}
\tag{3.24}
$$

The double occupancy of the orbitals ensures that such a wave function describe a singlet state, and the presence of only closed shells ensures that the wave function be totally symmetric.

The energy (3.10) of a closed-shell state reduces to

$$
E = 2\sum_{n=1}^{N/2} f_n + \sum_{m,n=1}^{N/2}(2J_{mn} - K_{mn}) + \sum_{\alpha > \beta = 1}^{\nu}\frac{Z_\alpha Z_\beta}{R_{\alpha\beta}}
\tag{3.25}
$$

in which the summations are now over the *orbitals*, with $f_n$ and $J_{mn}$ given by (3.11) and (3.12), and $K_{mn}$ by (3.13) without the integral over spin. Because the orbitals are associated with both spin factors, the operators $h_\alpha^{\mathrm{HF}}$ and $h_\beta^{\mathrm{HF}}$ are identical, and (3.23) can be replaced by the single set of (spin-independent) Hartree–Fock equations

$$
h^{\mathrm{RHF}}\psi_n = (f + 2\mathscr{J} - \mathscr{K})\psi_n = \epsilon_n\psi_n
\tag{3.26}
$$

where

$$
\mathscr{J}(1) = \sum_{m=1}^{N/2}\int\frac{\psi_m^*(2)\psi_m(2)}{r_{12}}\,dv_2
$$

$$
\mathscr{K}(1)\psi_n(1) = \sum_{m=1}^{N/2}\psi_m(1)\int\frac{\psi_m^*(2)\psi_n(2)}{r_{12}}\,dv_2
$$

The orbital energies are

$$
\epsilon_n = \int\psi_n^* h^{\mathrm{RHF}}\psi_n\,dv = f_n + \sum_{m=1}^{N/2}(2J_{mn} - K_{mn})
\tag{3.27}
$$

and the total energy can be written in the alternative forms

$$
\left.
\begin{aligned}
E &= 2\sum_{n=1}^{N/2}\epsilon_n - \sum_{m,n=1}^{N/2}(2J_{mn} - K_{mn}) + \sum_{\alpha>\beta=1}^{\nu}\frac{Z_\alpha Z_\beta}{R_{\alpha\beta}} \\
&= \sum_{n=1}^{N/2}(f_n + \epsilon_n) + \sum_{\alpha>\beta=1}^{\nu}\frac{Z_\alpha Z_\beta}{R_{\alpha\beta}}
\end{aligned}
\right\}
\tag{3.28}
$$

## The orbital approximation

We note that the total energy (apart from the constant nuclear repulsion terms) is not simply the sum of the orbital energies.

The eigenfunctions of the Hartree–Fock operator $h^{\mathrm{RHF}}$ form a complete set of orthonormal orbitals. An orbital approximation for the lowest totally symmetric singlet state is obtained when the $N/2$ orbitals which have the lowest (most negative) orbital energies are occupied. The corresponding orbital energies (3.27) are, since $J_{nn} = K_{nn}$,

$$\epsilon_n = f_n + J_{nn} + \sum_{\substack{m=1 \\ (m \neq n)}}^{N/2} (2J_{mn} - K_{mn}) \quad (n = 1, 2, ..., N/2) \qquad (3.29)$$

The Coulomb-exchange terms describe the interaction of an electron in the orbital $\psi_n$ with the other $(N-1)$ electrons; $J_{nn}$ for its interaction with the second electron in $\psi_n$, $2J_{mn}$ $(m \neq n)$ for its Coulomb interaction with the pair in $\psi_m$, and $-K_{mn}$ for its exchange interaction with that electron in $\psi_m$ which has the same spin. In addition to the occupied orbitals, there are the infinite number of unoccupied virtual orbitals $\psi_p$ $(p > N/2)$ whose orbital energies are

$$\epsilon_p = f_p + \sum_{m=1}^{N/2} (2J_{mp} - K_{mp}) \qquad (3.30)$$

In contrast to (3.29), the Coulomb-exchange terms describe the interaction of an electron in the virtual orbital $\psi_p$ with the $N$ electrons in the occupied orbitals $\psi_n$ $(n \leqslant N/2)$. The virtual orbitals therefore represent the states of a 'test' electron moving in the field of all the $N$ electrons, rather than one of the $N$ electrons moving in the field of the other $(N-1)$. These orbitals cannot therefore be regarded as suitable for the description of excited states of the system; a 'singly excited' configuration obtained by replacing one of the occupied spin-orbitals by a virtual spin-orbital is not normally a good approximation to an excited state.

*Example. The ground state of lithium hydride.* The $^1\Sigma^+$ ground state of LiH has the closed-shell wave function

$$\Psi = |1\sigma\alpha, 1\sigma\beta, 2\sigma\alpha, 2\sigma\beta|$$

with energy

$$E = 2f_{1\sigma} + 2f_{2\sigma} + J_{1\sigma,1\sigma} + J_{2\sigma,2\sigma} + 4J_{1\sigma,2\sigma} - 2K_{1\sigma,2\sigma} + 3/R_{\mathrm{LiH}}$$

The orbital energies of the occupied orbitals are

$$\epsilon_{1\sigma} = f_{1\sigma} + J_{1\sigma,1\sigma} + 2J_{1\sigma,2\sigma} - K_{1\sigma,2\sigma}$$

$$\epsilon_{2\sigma} = f_{2\sigma} + J_{2\sigma,2\sigma} + 2J_{1\sigma,2\sigma} - K_{1\sigma,2\sigma}$$

The energy if a virtual orbital, $3\sigma$ say, is

$$\epsilon_{3\sigma} = f_{3\sigma} + 2J_{1\sigma,3\sigma} + 2J_{2\sigma,3\sigma} - K_{1\sigma,3\sigma} - K_{2\sigma,\,3\sigma}$$

and represents the energy of an electron in the $3\sigma$ orbital in the presence of the four electrons of the neutral molecule.

### 3.3.2 Open-shell states

Although closed-shell states are necessarily totally symmetric singlet states, the example of the lowest $^1$S state of the carbon atom shows that the converse is not always true, and a wave function for this state cannot be expressed as a single Slater determinant. Thus, for the $(1s)^2(2s)^2(2p_m)^2$ orbital configuration of carbon, the single-determinant wave function

$$\Psi = |1s\alpha,\,1s\beta,\,2s\alpha,\,2s\beta,\,2p_m\alpha,\,2p_m\beta| \qquad (3.31)$$

is an orbital approximation for the lowest $^1$D state if $|m| = 1$, with quantum numbers $S = 0$, $L = 2$, and $|M_L| = 2$, but represents a mixture of the $^1$D and $^1$S states, with $M_L = 0$, if $m = 0$ (see table 2.6). The $^1$S state requires a combination of three Slater determinants for its representation (see (2.21) and (2.25)); the normalized wave function is

$$\Psi_{1_S} = (1/\sqrt{3})[\Psi_{0,0} - \Psi_{1,-1} - \Psi_{-1,1}] \qquad (3.32)$$

where $\qquad \Psi_{m,-m} = |1s\alpha,\,1s\beta,\,2s\alpha,\,2s\beta,\,2p_m\alpha,\,2p_{-m}\beta|$

Alternatively (3.32) can be written as

$$\Psi_{1_S} = (1/\sqrt{3})|1s\alpha,\,1s\beta,\,2s\alpha,\,2s\beta,\,[2p_0\alpha,\,2p_0\beta$$
$$- 2p_1\alpha,\,2p_{-1}\beta - 2p_{-1}\alpha,\,2p_1\beta]| \qquad (3.33)$$

The wave functions (3.31), with $|m| = 1$, and (3.33) are typical open-shell wave functions in the RHF scheme, containing a number of closed shells and (in this case) one open shell, the latter determining the symmetry properties of the wave function.

We can at this point consider a general definition of the RHF model. Given a basis of $M$ spin-orbitals $\phi_n$, it is possible to form a total of $\kappa = M!/N!(M-N)!$ independent Slater determinants $\Psi_I$ for an $N$-electron system, corresponding to the $\kappa$ ways of choosing $N$ from $M$ spin-orbitals. The most general wave function which can be formed from these, for any state, is the linear combination

$$\Psi = \sum_{I=1}^{\kappa} C_I \Psi_I \qquad (3.34)$$

## The orbital approximation

As the number of basis functions is increased indefinitely and the basis becomes complete, the set of Slater determinants also becomes complete, and the wave function can be made to converge to an exact eigenfunction of the Hamiltonian $\mathcal{H}$ of the system by a suitable choice of the coefficients $C_I$. The RHF description of a state of a system of $N$ electrons is that expansion (3.34) which contains the smallest possible number of Slater determinants consistent with the symmetry properties of the state in question. The corresponding orbitals are symmetry orbitals and each set of $l_i$ spatially degenerate orbitals belonging to symmetry species $\Gamma_i$ is associated with both $\alpha$- and $\beta$-spin factors to give a shell of $2l_i$ spin-orbitals. In addition, if the minimum number of Slater determinants required to represent an open-shell state is greater than one, the corresponding coefficients $C_I$, as in (3.32), are determined completely by symmetry and normalization.

For a large class of open-shell states, containing a single open shell, the energy can be expressed in the form

$$E = \{2\sum_k f_k + \sum_{k,\,l}(2J_{kl} - K_{kl})\} + \gamma\{2\sum_m f_m + \gamma\sum_{m,\,n}(2aJ_{mn} - bK_{mn})\}$$

$$+ 2\gamma\sum_k\sum_m(2J_{km} - K_{km}) + \sum_{\alpha>\beta}\sum\frac{Z_\alpha Z_\beta}{R_{\alpha\beta}} \quad (3.35)$$

The indices $k$ and $l$ refer to the closed-shell orbitals, and $m$ and $n$ to the open-shell orbitals. $\gamma$ is the 'fractional occupation number', $0 < \gamma < 1$, of the open shell, and is calculated by dividing the number of electrons in the shell by the total number of spin-orbitals which make up the shell. $a$ and $b$ are constants which depend on the coefficients in the symmetry-adapted combination (3.34) for the particular state being considered. The first group of terms in (3.35) is the energy of the closed shells alone, the second group is the energy of the open shell, and the third represents the interaction between these two.

Application of the variation principle to the energy (3.35), subject to the condition that the orbitals be orthonormal, gives the two sets of RHF equations (Roothaan 1960)

$$\left.\begin{array}{l} h_c^{\text{RHF}}\psi_k = \epsilon_k\psi_k \\ h_o^{\text{RHF}}\psi_m = \epsilon_m\psi_m \end{array}\right\} \quad (3.36)$$

in which $h_c^{\text{RHF}}$ is the Hartree–Fock operator for the closed-shell orbitals, and $h_o^{\text{RHF}}$ for the open-shell orbitals, both operators being totally symmetric. These operators, which are complicated functions of all the occupied orbitals and of the constants $\gamma$, $a$ and $b$, may be combined to give a single Hartree–Fock operator whose eigenfunctions include both the closed-shell and open-shell orbitals (McWeeny and Sutcliffe 1969, chapter 5).

### 3.4 KOOPMANS' THEOREM

Orbital energies are approximately observables in the sense that they provide estimates of the binding energies of the electrons in an atom or molecule, and these can be measured by the method of photoelectron spectroscopy (Turner 1968; Shirley 1973), The ionization process involving the ejection of an electron may be considered as taking place in two (fictitious) steps (Davies *et al.* 1970): (i) the photoelectron is ejected suddenly, leaving a 'hole' in the electron distribution, with the other (passive) electrons frozen in their initial ground-state orbitals, and (ii) the passive electrons relax towards the positive hole, accelerating the outgoing electron. The assumption of the first step alone as a description of the ionization makes up the 'sudden approximation' whereby, by Koopmans' theorem (1933), the binding energy is given by the orbital energy in the initial state. Koopmans' theorem is strictly valid for the unconstrained HF model and for closed-shell states in the RHF model. It is not generally valid for open-shell states and must then be used with some caution.

*Example. The ionization of lithium hydride.* The $^1\Sigma^+$ ground state of LiH (see the example on p. 50) has the closed-shell wave function

$$\Psi = |1\sigma\alpha, 1\sigma\beta, 2\sigma\alpha, 2\sigma\beta|$$

and energy

$$E = 2f_{1\sigma} + 2f_{2\sigma} + J_{1\sigma,1\sigma} + J_{2\sigma,2\sigma} + 4J_{1\sigma,2\sigma} - 2K_{1\sigma,2\sigma} + 3/R_{\text{LiH}}$$

The corresponding wave function and energy of the lowest $^2\Sigma^+$ state of the ion obtained by the 'sudden' removal of an electron from the $2\sigma$ orbital are

$$\Psi_+ = |1\sigma\alpha, 1\sigma\beta, 2\sigma\alpha|$$

$$E_+ = 2f_{1\sigma} + f_{2\sigma} + J_{1\sigma,1\sigma} + 2J_{1\sigma,2\sigma} - K_{1\sigma,2\sigma} + 3/R_{\text{LiH}}$$

and the ionization energy is therefore

$$E_+ - E = -[f_{2\sigma} + J_{2\sigma,2\sigma} + 2J_{1\sigma,2\sigma} - K_{1\sigma,2\sigma}] = -\epsilon_{2\sigma} \qquad (3.37)$$

It is clear from a consideration of the two-step description of the ionization process that the sudden approximation must overestimate the

## The orbital approximation

binding energy. Some results obtained from accurate RHF calculations are compared with experimental ionization energies in table 3.2. The agreement between the orbital energies and experimental (vertical) ionization energies is remarkably good, the error being normally not greater than 10 per cent, and sometimes much less. This is presumably due to an efficient cancellation of errors and although Koopmans' theorem can be, and frequently is, invoked for the interpretation of photoelectron spectra (see §6.6), it must be used with some caution, as the results for $N_2$ demonstrate. The ground-state configuration of $N_2$ is

$$^1\Sigma_g^+ : (1\sigma_g)^2 (1\sigma_u)^2 (2\sigma_g)^2 (2\sigma_u)^2 (3\sigma_g)^2 (1\pi_u)^4$$

and the orbital energies suggest that the $1\pi_u$ electrons are less strongly bound than the $3\sigma_g$ electrons. The ground state of the ion however has symmetry $^2\Sigma_g^+$, corresponding to the removal of an electron from the $3\sigma_g$ orbital of the neutral molecule. Reliable quantitative agreement of computed and experimental ionization energies in general requires not only separate orbital calculations for the neutral and ionic species, to allow for the relaxation of the electron distribution, but also separate treatments of electron correlation and, sometimes, separate estimates of relativistic corrections (see chapter 4). This is of course also true of the calculation of excitation energies. Although it is possible to derive simple expressions of the type (3.37) for excitation energies, we have already remarked that the individual virtual orbitals are not suitable for the description of excited states and cannot be expected to provide a reliable basis for the interpretation of electronic spectra.

TABLE 3.2  *Comparison of orbital energies and experimental ionization energies*

| Molecule | Reference | Orbital | $-\epsilon$ (RHF)/$H_\infty$ | I.E. (exp.)/$H_\infty$ |
|---|---|---|---|---|
| HCl | $a$ | $2\pi$ | 0.476 | 0.468 |
| $N_2$ | $b$ | $1\pi_u$ | 0.629 | 0.624 |
| | | $3\sigma_g$ | 0.638 | 0.580 |
| | | $2\sigma_u$ | 0.769 | 0.695 |
| $CH_4$ | $c$ | $1t_2$ | 0.546 | 0.529 |
| | | $2a_1$ | 0.944 | 0.840 |
| | | $1a_1$ | 11.20 | 10.68 |

References: (*a*) Cade and Huo 1967*b*; (*b*) Cade, Sales and Wahl 1966; (*c*) Meyer 1973.

## 3.5 BRILLOUIN'S THEOREM

The eigenfunctions of the Hartree–Fock operator ($h^{\text{HF}}$ or $h^{\text{RHF}}$) form a complete set of orthonormal functions, and it is therefore possible to form from them a complete set of $N$-electron Slater determinants, in terms of which an exact $N$-electron wave function may be expanded. Let the $N$ occupied spin-orbitals be labelled with the indices $k, l, m, \ldots$ ($= 1, 2, \ldots, N$) and the virtual spin-orbitals with the indices $p, q, r, \ldots$ ($> N$). The complete set of Slater determinants may then be considered as derived from the Hartree–Fock wave function

$$\Psi_{\text{HF}} = \Psi_0 = |\phi_1, \phi_2, \ldots, \phi_N|$$

by the substitution of sets of $n\, (= 1, 2, \ldots, N)$ occupied spin-orbitals by $n$ virtual spin-orbitals. For $n = 1$ we have a set of singly substituted ('excited') terms $\Psi_k^p$ obtained by replacing $\phi_k$ in $\Psi_0$ by $\phi_p$. For $n = 2$ we have a set of doubly substituted terms $\Psi_{kl}^{pq}$, and so on. Thus for a four-electron system,

$$\Psi_0 = |\phi_1, \phi_2, \phi_3, \phi_4|$$

and, for example,

$$\Psi_1^p = |\phi_p, \phi_2, \phi_3, \phi_4| \qquad \text{(all } p > 4)$$

$$\Psi_{12}^{pq} = |\phi_p, \phi_q, \phi_3, \phi_4| \qquad \text{(all } p > q > 4)$$

$$\Psi_{123}^{pqr} = |\phi_p, \phi_q, \phi_r, \phi_4| \qquad \text{(all } p > q > r > 4)$$

$$\Psi_{1234}^{pqrs} = |\phi_p, \phi_q, \phi_r, \phi_s| \qquad \text{(all } p > q > r > s > 4)$$

The determinants are normalized and orthogonal, and the exact wave function can now be expressed in the form

$$\Psi = C_0 \Psi_0 + \sum_k \sum_p C_k^p \Psi_k^p + \sum_{k<l} \sum_{p<q} C_{kl}^{pq} \Psi_{kl}^{pq} + \sum_{k<l<m} \sum_{p<q<r} C_{klm}^{pqr} \Psi_{klm}^{pqr}$$
$$+ \sum_{p<q<r<s} C_{1234}^{pqrs} \Psi_{1234}^{pqrs} \qquad (3.38)$$

for suitable values of the coefficients ($\sum_{k<l}$, for example, represents a double sum over the indices $k$ and $l$ with $k < l$). It is of course necessary to include only those terms consistent with the symmetry properties of the state in question.

Brillouin's theorem (1933) states that the matrix element of the Hamiltonian $\mathcal{H}$ between the Hartree–Fock wave function $\Psi_0$ and a singly substituted term $\Psi_k^p$ is zero:

$$\int \Psi_0^* \mathscr{H} \Psi_k^p d\tau = 0 \quad \text{(all } k \leqslant N \text{ and } p > N) \qquad (3.39)$$

A consequence of the theorem is that the singly substituted terms make no contribution to the wave function to 'first order', in a perturbation-theoretical sense (see §1.5), and are therefore expected to make only a small contribution to the energy. In addition, the total charge distribution calculated from the Hartree–Fock wave function is also correct to first order (see §6.1.2), and all those properties, such as the dipole moment, which depend solely on the charge distribution may be expected to be well described by the orbital approximation. Brillouin's theorem is strictly valid for the unconstrained HF model and for closed-shell states in the RHF model. It is not generally valid for open-shell states. The theorem is often invoked as a justification of the Hartree–Fock scheme and, although its usefulness depends greatly on how small a first-order quantity is in practice, it nevertheless provides a guide as to the type of property for which the Hartree–Fock wave function can be expected to give at least qualitative, and hopefully quantitative, agreement with experiment. This expectation is examined for the dipole moment in the following section.

### 3.5.1 *The dipole moment*

The classical dipole moment of a system of $\nu$ nuclei at positions $\boldsymbol{R}_\alpha$ and $N$ electrons at positions $\boldsymbol{r}_i$ is

$$\boldsymbol{\mu} = e \sum_{\alpha=1}^{\nu} Z_\alpha \boldsymbol{R}_\alpha - e \sum_{i=1}^{N} \boldsymbol{r}_i \qquad (3.40)$$

(independent of the position of the origin of the coordinate system if the molecule is electrically neutral). The value of the dipole moment in a state described by the normalized wave function $\Psi$ is the expectation (average) value (see §6.2)

$$\langle \boldsymbol{\mu} \rangle = \int \Psi^* \boldsymbol{\mu} \Psi \, d\tau = e \sum_{\alpha=1}^{\nu} Z_\alpha \boldsymbol{R}_\alpha - e \int \Psi^* \left( \sum_{i=1}^{N} \boldsymbol{r}_i \right) \Psi \, d\tau$$

For a single-determinant wave function,

$$\langle \boldsymbol{\mu} \rangle = e \sum_{\alpha=1}^{\nu} Z_\alpha \boldsymbol{R}_\alpha - e \sum_{n=1}^{N} \boldsymbol{\mu}_n$$

where $-e\mu_n$ is the contribution of an electron in the orbital $\psi_n$:

$$\mu_n = \int r\psi_n^* \psi_n \, dv$$

Dipole moments calculated from accurate RHF wave functions are compared in table 3.3 with the experimental values for several molecules which have closed-shell ground states. The table shows that the RHF model frequently gives better than qualitative agreement between computed and experimental dipole moments. The result for LiF is typical of the alkali halides, but in general an error of up to half a debye can be expected for the less polar molecules. The example of CO shows that the polarity may be described incorrectly within the orbital approximation for those molecules with very small dipole moments.

TABLE 3.3 *Computed (RHF) and experimental dipole moments*

| Molecule | Reference | $\mu$(RHF)/D | $\mu$(exp.)/D |
|----------|-----------|--------------|---------------|
| LiF | $a$ | 6.30 | 6.28 |
| CO | $a$ | 0.28 | $-0.11$ |
| HCN | $a$ | 3.29 | 2.95 |
| NNO | $a$ | 0.64 | 0·18 |
| FCN | $a$ | 2.28 | 2.17 |
| $H_2O$ | $b$ | 2.03 | 1.85 |
| $NH_3$ | $c$ | 1.66 | 1.48 |

The magnitudes are given in debyes, $D = 3.33564 \times 10^{-30}$ C m. For the linear molecules, a positive value means that the polarity is $+ -$ for the molecule as written.

References: ($a$) Yoshimine and McLean 1967; ($b$) Dunning, Pitzer and Aung 1972; ($c$) Rauk, Allen and Clementi 1970.

## 3.6 UNRESTRICTED AND EXTENDED HARTREE–FOCK MODELS

The restricted Hartree–Fock model, with its concept of a set of molecular (symmetry) orbitals each of which can be occupied by two electrons, forms the basis for much of the chemist's understanding and interpretation of the structure and properties of molecules. We saw in §3.2.2 however that the two simplifying and valuable properties of the orbitals, namely the double-occupancy and symmetry properties, in general represent constraints on the orbital approximation, and it follows that a lower, and therefore more accurate, energy can be obtained by a relaxation of these constraints.

There are at least two properties for which the RHF model can provide

# The orbital approximation

a quite incorrect description of the system. One is the Fermi contact interaction, or spin density at a nucleus, which is an important contribution to the hyperfine structure of atoms and molecules in open-shell states. An example is the $^4$S ground state of the nitrogen atom for which the RHF wave function is

$$\Psi_{RHF} = |1s\alpha, 1s\beta, 2s\alpha, 2s\beta, 2p_1\alpha, 2p_0\alpha, 2p_{-1}\alpha|$$

Only the $s$ orbitals make a non-zero contribution to the charge density at the nucleus, and the double occupancy of these orbitals means that there is no net electron spin at the nucleus. The RHF wave function therefore gives a zero Fermi contact term whereas the experimental value is non-zero. This deficiency of the model can be removed by relaxing the double-occupancy constraint to give the *spin-unrestricted* Hartree–Fock (SUHF) model (often called simply the unrestricted Hartree–Fock (UHF) model). Because of the unequal numbers of $\alpha$ and $\beta$ spins, the orbitals $\psi_n^{(\alpha)}$ associated with $\alpha$ spin satisfy a different eigenvalue equation from those, $\psi_n^{(\beta)}$, associated with $\beta$ spin. As discussed in §3.2.2, no orbital $\psi_n^{(\alpha)}$ is then in general the same as any orbital $\psi_n^{(\beta)}$. The SUHF wave function for the nitrogen atom is

$$\Psi_{SUHF} = |1s\alpha, 1s'\beta, 2s\alpha, 2s'\beta, 2p_1\alpha, 2p_0\alpha, 2p_{-1}\alpha|$$

with $1s \neq 1s'$ and $2s \neq 2s'$, and the contributions of the two types of orbitals to the spin density at the nucleus no longer cancel.

One objection to the SUHF model is that it produces wave functions that are not in general eigenfunctions of the spin operator $\mathscr{S}^2$. This is readily remedied by the application of a suitable projection operator to the computed wave function, in accordance with the discussion of §2.7, to give the *projected* Hartree–Fock (PHF) model:

$$\Psi_{PHF} = \mathscr{P}_S \Psi_{SUHF}$$

An alternative procedure is to form a symmetry-adapted wave function, for example in the form of a projected Slater determinant, *prior* to the calculation of the orbitals. This *extended* Hartree–Fock (EHF) model has the advantage that the resulting orbitals are the best orbitals consistent with the symmetry properties of the state in question.

The second property for which the RHF model can provide an incorrect description is the dissociation energy and the nature of the dissociation products. The RHF ground-state dissociation energy is defined as the difference between the sum of the RHF energies of the separated ground-state atoms and the RHF energy of the molecule in its

TABLE 3.4  *Computed (RHF) and experimental ground-state dissociation energies*

| Molecule | Reference | $D_e$ (RHF)/$H_\infty$ | $D_e$ (exp.)/$H_\infty$ |
|----------|-----------|------------------------|-------------------------|
| $H_2$ | *a* | 0.134 | 0.175 |
| LiH | *b* | 0.055 | 0.093 |
| HCl | *b* | 0.128 | 0.170 |
| $H_2O$ | *c* | 0.254 | 0.370 |
| $CH_4$ | *d* | 0.527 | 0.671 |
| LiF | *e* | 0.150 | 0.221 |
| CO | *e* | 0.291 | 0.413 |
| MgO | *e* | −0.037 | 0.140 |
| $CO_2$ | *e* | 0.415 | 0.619 |
| $N_2$ | *f* | 0.194 | 0.364 |
| $F_2$ | *g* | −0.050 | 0.062 |

References: (*a*) Kolos and Roothaan 1960; (*b*) Cade and Huo 1967*a*, *b*; (*c*) Dunning, Pitzer and Aung 1972; (*d*) Meyer 1973; (*e*) Yoshimine and McLean 1967; (*f*) Cade, Sales and Wahl 1966; (*g*) Wahl 1964.

(computed) equilibrium configuration:

$$D_e = \sum_{\text{atoms}} E_{\text{RHF}}(\text{atom}) - E_{\text{RHF}}(\text{molecule})$$

A number of dissociation energies obtained from accurate RHF calculations are compared in table 3.4 with the experimental values. It is clear from this selection that the RHF model does not provide a satisfactory basis for the prediction and interpretation of the process of molecular dissociation. The dissociation energies are generally much too small, and in some cases the negative values show that the RHF molecule is unstable with respect to dissociation into RHF atoms. One reason for this deficiency of the orbital model is of course that the dissociation energy of a molecule is a very small quantity compared with the total energies of the molecule and the separate atoms, and the errors in the total energies cannot be expected to cancel in general. For example, the RHF total and dissociation energies of $F_2$ are $-198.8H_\infty$ and $-0.05H_\infty$ respectively. An *additional* reason for the *restricted* orbital model arises from the double-occupancy and symmetry constraints which prevent the molecule from dissociating correctly into ground-state atoms. This may be illustrated by a consideration of the $^1\Sigma_g^+$ ground state of the hydrogen molecule, for which the RHF wave function is

$$\Psi = |1\sigma_g\alpha, 1\sigma_g\beta| = 1\sigma_g(1)\,1\sigma_g(2)\cdot(1/\sqrt{2})[\alpha(1)\beta(2)-\beta(1)\alpha(2)]$$

## The orbital approximation

A simple representation of the molecular orbital is a linear combination of $1s$ orbitals centred on the two nuclei (labelled A and B),

$$1\sigma_g = N[1s_A + 1s_B]$$

where $N$ is a normalization factor and, for example, $1s_A$ is a normalized hydrogen-like atomic orbital centred on A:

$$1s_A = (\zeta^3/\pi)^{\frac{1}{2}} e^{-\zeta r_A}$$

Application of the variation principle gives the dissociation energy $0.128H_\infty$, which is close to the RHF value of $0.134H_\infty$. The value of the parameter $\zeta$ at the computed equilibrium distance of $1.38a_\infty$ (the experimental distance is $1.40a_\infty$) is $1.2$, and this may be expected to decrease to a value close to its value of unity in the hydrogen atom as the internuclear separation increases. The *form* of the wave function however remains unchanged as the internuclear distance increases, and does not, in the limit of infinite separation of the nuclei, describe a system of two non-interacting neutral atoms. The spatial part of the wave function is

$$1\sigma_g(1)\,1\sigma_g(2) = N^2\{[1s_A(1)\,1s_B(2) + 1s_B(1)\,1s_A(2)]$$
$$+ [1s_A(1)\,1s_A(2) + 1s_B(1)\,1s_B(2)]\}$$

The first pair of terms, when $\zeta = 1$, is the correct (symmetrized) wave function for a pair of non-interacting hydrogen atoms, but the second pair is an approximate (symmetrized) wave function for the system $(H^+ + H^-)$. In the language of valence-bond theory, the covalent and ionic structures make equal contributions at all internuclear separations, and the RHF energy of the molecule approaches a value greater than the energy of $-1H_\infty$ of two neutral ground-state atoms, corresponding to dissociation products which are a mixture of ground and excited states of the systems $(H + H)$ and $(H^+ + H^-)$. This incorrect behaviour of the RHF potential energy curve is typical of those molecules in which the bonding, whether covalent or ionic, involves the creation of new electron pairs; that is, of most molecules. For example, the RHF dissociation products of LiF are $Li^+$ and $F^-$, and of CO are $C^+$ and $O^-$. In particular, this incorrect behaviour explains to a great extent the negative value of the dissociation energy obtained for $F_2$.

The conventional method of correcting the long-range behaviour is the method of *configuration interaction* (CI). To the $(1\sigma_g)^2$ configuration may be added the $(1\sigma_u)^2$ configuration, where

$$1\sigma_u = N'[1s_A - 1s_B]$$

The resulting wave function (excluding the singlet spin function) is

$$1\sigma_g(1)\,1\sigma_g(2) + \lambda 1\sigma_u(1)\,1\sigma_u(2)$$

where $\lambda$ is a variational parameter. If the same atomic orbitals are used in both molecular orbitals, the wave function is, on expansion,

$$[1s_A(1)\,1s_B(2) + 1s_B(1)\,1s_A(2)] + C[1s_A(1)\,1s_A(2) + 1s_B(1)\,1s_B(2)]$$

and the ratio of the contributions of covalent and ionic structures is now a variationally determined function of the internuclear distance, and $C$ vanishes as the molecule dissociates correctly into ground-state atoms. This calculation was carried out by Weinbaum (1933) and gives a dissociation energy of $0.148H_\infty$. A further improvement is obtained when other parameters are included, and the orbitals allowed to vary independently.

Although a CI formulation has been used for the Weinbaum wave function, it is easy to show that an extended Hartree–Fock (EHF) formulation gives the same results, if both the double-occupancy and symmetry constraints are relaxed. The Weinbaum wave function can be written in the EHF form

$$\Psi_{\mathrm{EHF}} = N\{|\sigma\alpha, \sigma'\beta| + |\sigma'\alpha, \sigma\beta|\}$$

in which the orbitals are

$$\sigma = 1s_A + a1s_B, \quad \sigma' = 1s_B + a1s_A$$

This wave function may be obtained from either single determinant by the application of the spin projection operator $\mathscr{P}_S$ for $S = 0$. Expansion of the wave function gives

$$\Psi_{\mathrm{EHF}} = N[\sigma(1)\,\sigma'(2) + \sigma'(1)\,\sigma(2)] \cdot (1/\sqrt{2})\,[\alpha(1)\,\beta(2) - \beta(1)\alpha(2)]$$

and this is identical to the Weinbaum wave function if

$$C = 2a/(1 + a^2)$$

It is clear from the above discussion that not only does the EHF model provide a link between the traditional molecular-orbital and valence-bond theories, but it forms a basis for the description of molecules which can be considerably more realistic and accurate than that of the RHF model. We have seen on the other hand that one alternative way of proceeding beyond the RHF model is by the method of configuration interaction. In addition, CI is an example of the general method of linear combinations discussed in §1.3, and it provides, unlike the EHF method, one way of obtaining solutions of the Schrödinger equation of any desired accuracy.

# 4 Beyond the orbital approximation

## 4.1 ELECTRON CORRELATION

The orbital approximation is the starting point for several of the methods that have been devised for the calculation of accurate solutions of the Schrödinger equation. An eigenvalue of the non-relativistic Hamiltonian (1.6) of an isolated system is then conveniently written as the sum of two contributions:

$$E = E_{HF} + E_C$$

in which $E_{HF}$ is the Hartree–Fock energy and $E_C$, which corrects for the deficiencies of the orbital approximation, is called the *electron correlation* energy. The definition of $E_C$ of course depends on the particular Hartree–Fock model used to calculate $E_{HF}$. The correlation energy is smallest if $E_{HF}$ is calculated for the completely unconstrained model or, if the HF wave function is to have the correct symmetry properties, for the extended model. The convention that has evolved is that, because the great majority of orbital calculations are performed within the restricted Hartree–Fock scheme, the correlation energy is defined as the difference between the exact non-relativistic energy and the RHF energy (Löwdin 1959a):

$$E = E_{RHF} + E_C \qquad (4.1)$$

As a result, $E_C$ contains contributions which, as we have seen in the previous section, could be obtained within the orbital approximation by the relaxation of the double-occupancy and symmetry constraints. These contributions can be a substantial part of the correlation energy, and a consideration of the differences between the RHF model and, for example, the EHF model can be used to shed considerable light on the nature of electron correlation.

We consider the $^1$S ground state of the helium atom, for which the RHF wave function is

$$\Psi_{RHF} = |1s\alpha, 1s\beta|$$

$$= 1s(1)\,1s(2)\cdot(1/\sqrt{2})\,[\alpha(1)\,\beta(2) - \beta(1)\,\alpha(2)] \qquad (4.2)$$

Both electrons occupy the same $1s$ orbital, and the probability of finding either electron at position $r$ is $1s(r)^2$, independent of the position of the other electron. This is clearly not a correct description since the Coulomb repulsion between the electrons requires that the probability of finding one electron at $r$ should decrease as the other electron approaches this position. The electrostatic interactions of electrons therefore create a *Coulomb hole* (Löwdin 1959a) around each electron within which the probability of finding any other electron is diminished. Two effects can be distinguished for the helium atom. The first is *radial correlation*, whereby the electrons have a tendency to be at different distances from the nucleus. Part of the radial correlation energy can be obtained by a relaxation of the double-occupancy constraint and allowing the electrons to occupy different $1s$ orbitals; that is, by replacing the orbital product in (4.2) by the (symmetrized) product

$$1s(1)\,1s'(2) + 1s'(1)\,1s(2) \qquad (4.3)$$

A simple wave function of this form is obtained by putting (Hylleraas 1929)

$$1s = e^{-\zeta_1 r}, \quad 1s' = e^{-\zeta_2 r}$$

Application of the variation principle gives the parameter values $\zeta_1 = 1.19$ and $\zeta_2 = 2.18$, and an energy $-2.876 H_\infty$ which is appreciably lower than the RHF limit of $-2.862 H_\infty$ (see table 3.1). The values of the parameters show that in this 'split-shell' description of the atom an electron in the $1s$ orbital is on average twice as far from the nucleus as an electron in the $1s'$orbital.

The second effect in helium is *angular correlation*, whereby the electrons have a tendency to be on opposite sides of the nucleus This may again be described within the EHF scheme, this time by a relaxation of the symmetry constraint as well as of the double-occupancy constraint. A simple way of separating the electrons is in terms of the $sp$-type 'hybrid' orbitals

$$\psi = 1s + C2p_z, \quad \psi' = 1s' - C2p_z$$

directed towards the positive and negative $z$ directions respectively. The corresponding extended orbital approximation is (apart from normalization)

$$\mathscr{P}_L\mathscr{P}_S|\psi\alpha, \psi'\beta| \quad (L = S = 0) \qquad (4.4)$$

where $\mathscr{P}_L$ and $\mathscr{P}_S$ are orbital and spin symmetry projection operators, and the spatial factor of this wave function is

$$1s(1)\,1s'(2) + 1s'(1)\,1s(2)$$
$$-\tfrac{2}{3}C^2[2p_x(1)\,2p_x(2) + 2p_y(1)\,2p_y(2) + 2p_z(1)\,2p_z(2)] \qquad (4.5)$$

If, for example, $2p_x = xe^{-\zeta r}$, (4.5) reduces to (see fig. 4.1)

$$1s(1)\,1s'(2) + 1s'(1)\,1s(2) - \tfrac{2}{3}C^2 r_1 r_2 e^{-\zeta(r_1+r_2)}\cos\theta$$

63

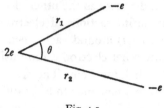

Fig. 4.1

For given values of $r_1$ and $r_2$, this wave function has its maximum value when $\theta = 180°$, with the electrons on opposite sides of the nucleus, and has its minimum value when $\theta = 0$, with the electrons on the same side of the nucleus. Such a wave function, with $C$ and $\zeta$ as variational parameters, gives a ground-state energy of $-2.895H_\infty$ (Taylor and Parr 1952). The exact ground-state energy is $-2.9037H_\infty$, so that nearly 80 per cent of the correlation energy is accounted for by this simple extended orbital approximation. The most accurate EHF calculation has produced an energy $-2.9028H_\infty$ (Bunge 1967) for a wave function of type (4.4), and $-2.9036H_\infty$ when the spin constraint on the spin-orbitals (see the footnote to p. 37) is also relaxed.

It may be concluded from the preceding results that a large part of the correlation energy has its origins in the double-occupancy and symmetry constraints imposed on the orbitals of the RHF model, and that a highly accurate, and still physically simple and appealing, description of an $N$-electron system is provided by the EHF model. On the other hand, the EHF wave function cannot be an exact solution of the Schrödinger equation and, as the wave function (4.5) implies, the EHF method is a type of CI method, in which the expansion coefficients are constrained by the original single-determinant form. The CI method is the more general, and computationally simpler.

Whereas, as in the ground state of the helium atom, the motions of electrons with unlike spins are completely uncorrelated in the RHF model, satisfaction of the Pauli principle already correlates to a great extent the motions of electrons with like spins. The Pauli principle creates a *Fermi hole* around each electron within which the probability of finding any other electron with the same spin is small, and the wave function vanishes when two electrons with like spins are at the same position in space. As an example, the lowest ³S state of the helium atom may be described by the single determinant (for $M_S = 1$)

$$|1s\alpha, 2s\alpha| = (1/\sqrt{2})[1s(1)\,2s(2) - 2s(1)\,1s(2)]\,\alpha(1)\,\alpha(2)$$

and the spatial factor vanishes when the spatial coordinates of the two electrons are the same. The greater part of the correlation energy is therefore associated with electrons with unlike spins. In the orbital approximation, two electrons are on average closer to each other when they occupy the same orbital than when they occupy different orbitals. The correlation energy of a pair of electrons in the same orbital (*intrapair correlation*) is consequently larger in general than that of two electrons in different orbitals (*interpair correlation*). This should be particularly true for localized orbitals. The correlation contribution to, for example, the energy of formation of a molecule can then be understood in terms of the creation of new electron pairs, and the resulting proximity of previously separated electrons. This is also true of the energy changes that accompany excitations and ionizations, particularly when these processes involve the destruction or creation of electron pairs. Although the correlation energy is always a small fraction (less than 1 per cent) of the total electronic energy of the system, it is nevertheless usually very much larger than the energy changes that are of interest in chemistry. For this reason, calculations of such energy changes often require the inclusion of at least a limited amount of CI in order to balance the accuracies of the energies of the various states involved.

## 4.2 CONFIGURATION INTERACTION

We have seen how an exact wave function for an $N$-electron system can be expressed as a linear combination of a complete set of Slater determinants. Each Slater determinant is a *spin-orbital configuration* and represents a particular distribution of the $N$ electrons amongst the available spin-orbitals (it is also often called an orbital configuration or an electron configuration). The Slater determinants may not however in general have the correct space and spin symmetry properties of the state in question, and may then make no contribution to the wave function. It is therefore sometimes more convenient to replace the individual Slater determinants by symmetry-adapted linear combinations of determinants, which can be constructed, for example, by the use of projection operators. Each such symmetry-adapted combination is called a *configuration*, and only those configurations with the correct symmetry properties of the state in question need be included in the expansion of the wave function. As an example, for the helium atom, the spin-orbital configurations

$$|1s\alpha, 2s\beta| \quad \text{and} \quad |1s\beta, 2s\alpha|$$

## Beyond the orbital approximation

can be transformed into the configurations

$$(1/\sqrt{2})[|1s\alpha, 2s\beta| + |1s\beta, 2s\alpha|] \quad \text{and} \quad (1/\sqrt{2})[|1s\alpha, 2s\beta| - |1s\beta, 2s\alpha|]$$

of which the first has $^1$S symmetry and the second has $^3$S symmetry. They both represent the *orbital configuration* $(1s)(2s)$, but with different spin couplings.

Denoting the set of Slater determinants (or of configurations) by $\Psi_I$ $(I = 0, 1, 2, ...)$ the exact wave function for any state is

$$\Psi = \sum_I C_I \Psi_I$$

If the orbitals, and therefore the $\Psi_I$, are fixed known functions, then the calculation of the wave function is reduced to the solution of a secular problem, as discussed in §1.3. In practice, of course, it is not normally possible to solve a matrix eigenvalue problem of infinite dimension, even if a complete set of functions is available. From a set of $M$ spin-orbitals it is possible to form a total $M!/N!(M-N)!$ independent $N$-electron configurations and, although not all of these will have the correct symmetry properties of the state in question, this number very quickly becomes too large for a complete CI calculation to be a practical proposition. Thus for 10 electrons and 20 orbitals (40 spin-orbitals), there are nearly $10^9$ configurations. It is therefore of some importance not only to be able to choose a set of orbitals for which the CI expansion converges rapidly, but also to be able to predict which configurations make significant contributions to the wave function and must be included in the CI expansion.

The various methods that have been devised to solve the convergence problem are essentially of two types, depending on whether the choice of orbitals precedes and determines the choice of configurations or the choice of configurations precedes and determines the choice of orbitals. In the former case, the configurations are built up from some previously chosen fixed set of one-electron functions which may be, but are not necessarily, the Hartree–Fock orbitals for the state in question. Such an approach almost inevitably leads to a slowly convergent CI expansion unless the orbitals are very carefully chosen, with many configurations making small but significant contributions to the wave function. As will be discussed in the following section, the choice of the important configurations is simplified somewhat if the orbitals are the Hartree–Fock orbitals. The alternative approach is to choose first a set of configurations which, on qualitative grounds based on considerations such as those of the previous section and of §4.3, can be expected to account for the bulk of the correlation energy; for example, all those configurations which are necessary for the correct description of the dissociation of a molecule. The orbitals from which these configurations are constructed are then calculated to minimize the energy. The advantage of this over the fixed-orbital approach is that far fewer configurations are normally required to produce a given accuracy (tens or hundreds instead of hundreds or thousands). For a given set of configurations the optimum orbitals may be obtained by the *multiconfigurational SCF* method (MC–SCF) (Veillard and Clementi 1967; Das and Wahl 1972a; McWeeny and Sutcliffe 1969, §5.6). The many-configuration wave function is treated in the same way as the single-configuration wave function of Hartree–Fock theory, and the variation principle is used to derive a set of one-electron eigenvalue equations, analogous to the Hartree–Fock equations, which can be solved by the SCF procedure.

Several procedures have been devised for the practical implementation of the CI method (for a useful survey of these, see Schaefer 1972), and they have two features in common. Firstly, they are all very costly, and any saving of computational labour in one part of the calculation is usually balanced by increased labour in another part. Secondly, the resulting wave functions are not readily interpreted in terms of familiar physical pictures. The question of interpretation is discussed in the following section and in chapter 6. Accurate molecular CI calculations, yielding 80 per cent or more of the correlation energy, have, at the time of writing, been performed only for diatomic molecules and some simple polyatomic molecules such as water and methane. They have nevertheless already provided a considerable quantitative as well as qualitative insight into the nature of electron correlation and the changes that occur in a molecule on dissociation, excitation and ionization. Two such calculations, for the fluorine and methane molecules, are discussed in §4.4.

**4.3** PERTURBATIVE TREATMENT OF ELECTRON CORRELATION

In a CI expansion in which the Slater determinants are constructed from Hartree–Fock orbitals (restricted or more general), the Hartree–Fock wave function often makes by far the largest single contribution to the wave function, particularly for closed-shell ground states. This suggests that one way of proceeding beyond the orbital approximation is by the method of perturbation theory, whereby the difference between the

67

## Beyond the orbital approximation

'exact' model (as represented by the Hamiltonian $\mathscr{H}$) and the Hartree–Fock model is treated as a perturbation on the Hartree–Fock model (Sinanoglu 1964; McWeeny and Steiner 1965).

We consider, for simplicity, a state for which the RHF wave function is a single Slater determinant,

$$\Psi_{\mathrm{RHF}} = |\phi_1, \phi_2, ..., \phi_N|$$

and the orbitals are eigenfunctions of the one-electron Hartree–Fock operator $h^{\mathrm{RHF}}$;

$$h^{\mathrm{RHF}} \psi_n = \epsilon_n \psi_n$$

The total wave function is then an eigenfunction of the $N$-electron Hartree–Fock operator

$$\mathscr{H}^{\mathrm{RHF}} = \sum_{i=1}^{N} h^{\mathrm{RHF}}(i)$$

with eigenvalue equal to the sum of the orbital energies:

$$E^{\mathrm{RHF}} = \sum_{n=1}^{N} \epsilon_n$$

More generally, any Slater determinant of $N$ Hartree–Fock spin-orbitals is an eigenfunction of $\mathscr{H}^{\mathrm{RHF}}$ with eigenvalue equal to the sum of the corresponding orbital energies.

The total Hamiltonian of the system can be written as

$$\mathscr{H} = \mathscr{H}^{\mathrm{RHF}} + \mathscr{H}_1$$

in which $\mathscr{H}^{\mathrm{RHF}}$ may be identified with the unperturbed Hamiltonian $\mathscr{H}_0$ of (1.15) in §1.5, and $\mathscr{H}_1$, identified with $\lambda V$ in (1.15), is a 'correlation potential' which is treated as the perturbation. Following the procedure outlined in §1.5, we expand the energy and wave function for the (perturbed) system as power series in the perturbation:

$$E_0 = E_0^{(0)} + E_0^{(1)} + E_0^{(2)} + ...$$

$$\Psi_0 = \Psi_0^{(0)} + \Psi_0^{(1)} + \Psi_0^{(2)} + ...$$

in which the state of interest has been labelled with the index o, so that

$$\Psi_0^{(0)} = \Psi_{\mathrm{RHF}}, \quad E_0^{(0)} = E^{\mathrm{RHF}}$$

Then

$$E_0^{(1)} = \int \Psi_{\mathrm{RHF}}^* \mathscr{H}_1 \Psi_{\mathrm{RHF}} \, \mathrm{d}\tau$$

and

$$E_0^{(0)} + E_0^{(1)} = \int \Psi_{\mathrm{RHF}}^* \mathscr{H} \Psi_{\mathrm{RHF}} \, \mathrm{d}\tau$$

is the total energy ($E_{RHF}$ of (4.1)) of the system in the RHF approximation. The second-order energy $E_0^{(2)}$ is therefore the first, and dominant, term in the perturbative expansion of the correlation energy $E_C$.

The first-order wave function $\Psi_0^{(1)}$ can be expanded as a linear combination of the complete set of unperturbed eigenfunctions (the Slater determinants) $\Psi_I^{(0)}$ in accordance with (1.26):

$$\Psi_0^{(1)} = \sum_{I \neq 0} \left( \frac{H_{I0}}{E_0^{(0)} - E_I^{(0)}} \right) \Psi_I^{(0)} \tag{4.6}$$

where

$$H_{I0} = \int \Psi_I^{(0)*} \mathscr{H} \Psi_{RHF} \, d\tau$$

Let the $N$ occupied spin-orbitals in the Hartree–Fock wave function $\Psi_{RHF}$ be labelled with the indices $k, l, m \ldots (= 1, 2, \ldots, N)$ and the virtual spin-orbitals with the indices $p, q, r, \ldots (> N)$. As discussed in §3.5, the Slater determinants $\Psi_I^{(0)}$ ($I \neq 0$) can be expressed in the form $\Psi_{klm\ldots}^{pqr\ldots}$, in which 1 to $N$ occupied spin-orbitals $\phi_k, \phi_l, \phi_m, \ldots$ in $\Psi_{RHF}$ have been replaced by the virtual spin-orbitals $\phi_p, \phi_q, \phi_r, \ldots$. The first-order wave function (4.6) is then (see (3.38))

$$\Psi_0^{(1)} = \sum_k \sum_p C_k^p \Psi_k^p + \sum_{k<l} \sum_{p<q} C_{kl}^{pq} \Psi_{kl}^{pq} + \ldots \tag{4.7}$$

where

$$C_{kl\ldots}^{pq\ldots} = \frac{\int \Psi_{kl\ldots}^{pq\ldots *} \mathscr{H} \Psi_{RHF} \, d\tau}{(\epsilon_k + \epsilon_l + \ldots) - (\epsilon_p + \epsilon_q + \ldots)} \tag{4.8}$$

and the corresponding second-order energy is

$$E_0^{(2)} = \int \Psi_{RHF}^* \mathscr{H}_1 \Psi_0^{(1)} \, d\tau = \int \Psi_{RHF}^* \mathscr{H} \Psi_0^{(1)} \, d\tau$$

$$= \sum_k \sum_p C_k^p \int \Psi_{RHF}^* \mathscr{H} \Psi_k^p \, d\tau + \sum_{k<l} \sum_{p<q} C_{kl}^{pq} \int \Psi_{RHF}^* \mathscr{H} \Psi_{kl}^{pq} \, d\tau + \ldots \tag{4.9}$$

It can be shown that, since the Hamiltonian $\mathscr{H}$ is a sum of one- and two-electron operators only, a matrix element

$$H_{IJ} = \int \Psi_I^{(0)*} \mathscr{H} \Psi_J^{(0)} \, d\tau$$

is zero if the determinants $\Psi_I^{(0)}$ and $\Psi_J^{(0)}$ differ in more than two spin-orbitals. In particular, $H_{I0}$ is zero if $\Psi_I^{(0)}$ contains more than two virtual spin-orbitals. The expansions (4.7) and (4.9) are therefore truncated after

## Beyond the orbital approximation

the first two sets of terms, and contain contributions from the singly and doubly substituted configurations only. This is a general property of the first-order wave function. In addition, when the orbitals are Hartree–Fock orbitals and when Brillouin's theorem (§3.5) applies, (3.39) shows that the singly substituted configurations also vanish. The first-order wave function then contains only doubly substituted configurations,

$$\Psi_0^{\prime(1)} = \sum_{k<l} \sum_{p<q} C_{kl}^{pq} \Psi_{kl}^{pq} \tag{4.10}$$

and the second-order energy is

$$E_0^{(2)} = \sum_{k<l} \sum_{p<q} C_{kl}^{pq} \int \Psi_{\mathrm{RHF}}^* \mathscr{H} \Psi_{kl}^{pq} \, d\tau \tag{4.11}$$

Like Brillouin's theorem, the expressions (4.10) and (4.11) are valid for the unconstrained HF model and for closed-shell states in the RHF model. The singly substituted configurations must in general also be included when Brillouin's theorem fails for open-shell states.

The foregoing analysis shows that the doubly substituted configurations, and the singly substituted configurations for open-shell states, can in general be expected to make the most important contributions to the correlation energy. This provides a first rule for the choice of configurations in a CI expansion of the wave function, even when the orbitals are not Hartree–Fock orbitals.

### 4.3.1 Pair-correlated wave functions

Perhaps the most significant aspect of the perturbative treatment of electron correlation is not that it provides in itself a tractable method of calculating correlation energies, but that it has generated a language and a set of physical concepts in terms of which the electron correlation in an $N$-electron system may be better understood. We note that the first-order wave function (4.10) can be written in the form (dropping the subscript o)

$$\Psi^{\prime(1)} = \sum_{k<l} \Phi_{kl}^{(1)}$$

where

$$\Phi_{kl}^{(1)} = \sum_{p<q} C_{kl}^{pq} \Psi_{kl}^{pq} \tag{4.12}$$

$\Phi_{kl}^{(1)}$ is a *pair-correlated* wave function, and is a combination of those Slater determinants that are obtained from $\Psi_{\mathrm{RHF}}$ by the replacement of the pair of 'occupied' spin-orbitals $(\phi_k, \phi_l)$ by all possible pairs of virtual

70

spin-orbitals $(\phi_p, \phi_q)$. As an example, the RHF wave function for the ground state of beryllium is

$$\Psi_{\text{RHF}} = |1s\alpha, 1s\beta, 2s\alpha, 2s\beta|$$

$$= (4!)^{-\frac{1}{2}} \det [1s(1)\alpha(1)\, 1s(2)\,\beta(2)\, 2s(3)\,\alpha(3)\, 2s(4)\,\beta(4)]$$

and one of the pair-correlated wave functions (4.12) is

$$\Phi^{(1)}_{1s\alpha,\, 1s\beta} = (4!)^{-\frac{1}{2}} \sum_{p,\,q} C^{p\alpha,\, q\beta}_{1s\alpha,\, 1s\beta} \det [\psi_p(1)\,\alpha(1)\,\psi_q(2)\,\beta(2)\, 2s(3)\,\alpha(3)\, 2s(4)\,\beta(4)]$$

$$= (4!)^{-\frac{1}{2}} \det [\phi^{(1)}_{1s\alpha,\, 1s\beta}(1, 2)\, 2s(3)\,\alpha(3)\, 2s(4)\,\beta(4)]$$

where $\qquad \phi^{(1)}_{1s\alpha,\, 1s\beta}(1, 2) = \sum_{p,\,q} C^{p\alpha,\, q\beta}_{1s\alpha,\, 1s\beta} \psi_p(1)\,\psi_q(2)\,\alpha(1)\,\beta(2)$ $\qquad(4.13)$

and the summation is over all pairs of virtual orbitals. $\phi^{(1)}_{1s\alpha,\, 1s\beta}$ is called a *correlated pair function* and it describes, to first order, the *intrapair* correlation between the motions of the two electrons in the $1s$ orbital. There is a similar function for the pair of electrons in the $2s$ orbital, and a set of four pair functions which describe the *interpair* correlation between the motion of an electron in the $1s$ orbital and an electron in the $2s$ orbital; for example,

$$\Phi^{(1)}_{1s\alpha,\, 2s\alpha} = (4!)^{-\frac{1}{2}} \det [\phi^{(1)}_{1s\alpha,\, 2s\alpha}(1, 3)\, 1s(2)\,\beta(2)\, 2s(4)\,\beta(4)]$$

and $\qquad \phi^{(1)}_{1s\alpha,\, 2s\alpha}(1, 2) = \sum_{p,\,q} C^{p\alpha,\, q\alpha}_{1s\alpha,\, 2s\alpha} \psi_p(1)\,\psi_q(2)\,\alpha(1)\,\alpha(2)$

The correlation energy, as represented by the second-order energy (4.11), therefore separates into a sum of *pair-correlation energies*,

$$E_C \simeq E^{(2)} = \sum_{k<l} e^{(2)}(k, l) \qquad (4.14)$$

where $\qquad e^{(2)}(k, l) = \int \Psi^*_{\text{RHF}} \mathscr{H} \Phi^{(1)}_{kl}\, d\tau \qquad (4.15)$

is the correlation energy associated with the electrons in the pair of spin-orbitals $\phi_k$ and $\phi_l$.

In general, the correlation between the motions of a pair of electrons in spin-orbitals $\phi_k$ and $\phi_l$ is described, in first order, by replacing the corresponding product in $\Psi_{\text{RHF}}$ by the correlated pair function $\phi^{(1)}_{kl}$:

$$\phi_k(1)\,\phi_l(2)\,\phi_m(3) \cdots \rightarrow \phi^{(1)}_{kl}(1, 2)\,\phi_m(3) \cdots$$

The expansion (4.13) in terms of the virtual spin-orbitals is only one of many possible representations of a pair function; for example, the various methods that have been developed for the solution of the Schrödinger

## Beyond the orbital approximation

equation of the helium atom (Hylleraas 1964) and of the hydrogen molecule (Slater 1963, chapters 3 and 4) may be adapted for use in the present context, and the first-order wave function calculated from the variational form (1.28) of the second-order energy.

The result that the correlation energy can often be approximated as the sum of pair energies has led to the development of several methods of calculating the pair functions and pair energies directly and independently (Nesbet 1969; Sinanoglu 1969; Schaefer 1972). For example, if the wave function is assumed to be of the pair-correlated form

$$\Psi = \Psi_{\mathrm{RHF}} + \sum_{k<l} \Phi_{kl} \tag{4.16}$$

in which $\Phi_{kl}$ contains a correlated pair function $\phi_{kl}$, then one way of calculating approximate pair functions is to apply the variation principle to a wave function

$$\Psi_{kl} = \Psi_{\mathrm{RHF}} + \Phi_{kl}$$

for each pair. The corresponding energy is

$$E_{kl} = \int \Psi_{kl}^{*} \mathscr{H} \Psi_{kl} \, d\tau \Big/ \int \Psi_{kl}^{*} \Psi_{kl} \, d\tau$$

$$= E_{\mathrm{RHF}} + e(k, l) \tag{4.17}$$

and the total correlation energy can be approximated as

$$E_{\mathrm{C}} \simeq \sum_{k<l} e(k, l) \tag{4.18}$$

Results obtained by Nesbet (1967) for the ground state of the beryllium atom are shown in table 4.1.

A second consequence of the observation that the correlation energy

TABLE 4.1  *Pair correlation energies in the ground state of beryllium according to Nesbet (1967)*

| $k$ | $l$ | $-e(k, l)/H_{\infty}$ |
|---|---|---|
| $1s\alpha$ | $1s\beta$ | 0.0418 |
| $1s\alpha$ | $2s\alpha$ | 0.0008 |
| $1s\alpha$ | $2s\beta$ | 0.0021 |
| $1s\beta$ | $2s\alpha$ | 0.0021 |
| $1s\beta$ | $2s\beta$ | 0.0008 |
| $2s\alpha$ | $2s\beta$ | 0.0454 |
| Total intrapair energy | | $-0.0872H_{\infty}$ |
| Total interpair energy | | $-0.0058H_{\infty}$ |
| Total pair energy | | $-0.0930H_{\infty}$ |
| Exp. correlation energy | | $-0.094H_{\infty}$ |

may be approximated as a sum of pair energies is that it provides a method of analysing the energy obtained, for example, by an extensive CI calculation. Thus, for a CI wave function which can be written as

$$\Psi = \Psi_{RHF} + \sum_{k<l} \Phi_{kl} + \text{smaller terms}$$

the pair energies can be obtained either from (4.17) or from an expression like (4.15). An example of such an analysis for the methane molecule is given in the next section.

**4.4** ELECTRON CORRELATION IN METHANE AND FLUORINE

The ground states of the methane and fluorine molecules are closed-shell states, and the foregoing discussion suggests that an accurate description of such states is provided by pair-correlated wave functions of the form (4.16). The calculations described here are essentially of this type although they differ greatly in the way the pair functions are obtained.

*The ground state of methane* (Meyer 1973). The ground state of the methane molecule has the RHF orbital configuration

$$(1a_1)^2(2a_1)^2(1t_2)^6$$

The results which are summarized in table 4.2 have been obtained from a CI expansion made up of 230 configurations formed from a total of

TABLE 4.2 *Total, correlation, dissociation and vertical ionization energies for the ground state of methane* (*Meyer 1973*)

|  | RHF | CI | Exp. |
|---|---|---|---|
| $E$ | $-40.214$ | $-40.458$ | $-40.515$ |
| $E_C$ |  | $-0.244$ | $-0.295$† |
| $D_e$ | 0.527 | 0.636 | 0.671 |

| | VERTICAL IONIZATION ENERGIES | | | |
|---|---|---|---|---|
| | Koopmans | RHF‡ | CI‡ | Exp. |
| $1a_1$ | 11.20 | 10.69 | 10.68 | 10.68 |
| $2a_1$ | 0.944 | 0.893 | 0.870 | 0.840 |
| $1t_2$ | 0.546 | 0.502 | 0.523 | 0.529 |

All energies (in units of $H_\infty$) are for the experimental bond length $2.05a_\infty$.

† Experimental correlation energy obtained from an estimated RHF limiting energy of $-40.22H_\infty$.

‡ Obtained from RHF and CI calculations for the appropriate states of $CH_4^+$, at the ground-state geometry and in terms of the same set of orbitals as the corresponding ground-state calculation.

TABLE 4.3  *Pair-correlation energies in the ground state of methane*
*(Meyer 1973)*

| Orbital pairs | Pair energies/$H_\infty$ |
|---|---|
| $(1a_1)^2$ | 0.0363 (1) |
| $(1a_1, 2a_1)$ | 0.0032 (4) |
| $(1a_1, 1t_2)$ | 0.0075 (12) |
| $(2a_1)^2$ | 0.0095 (1) |
| $(2a_1, 1t_2)$ | 0.0602 (12) |
| $(1t_2)^2$ | 0.0459 (3) |
| $(1t_2, 1t_2')$ | 0.0817 (12) |
| | |
| Total pair energy | $0.2443H_\infty$ |
| Total intrapair energy | $0.0917H_\infty$ |
| Total interpair energy | $0.1526H_\infty$ |
| Inter/intra = 1.67 | |

The figures in brackets are the numbers of spin-orbital pairs corresponding to the same orbital pair.

356 orbitals. Although the calculation is of the fixed-orbital type discussed in §4.2, the high accuracy of the results is due to a careful choice of orbitals, with a different appropriately selected set for each pair function. In table 4.3 the correlation energy is analysed in terms of pair energies of the distinct orbital pairs. The total pair energy can be written as

$$E_C = E_C(\text{core}) + E_C(\text{valence}) + E_C(\text{intershell})$$

$E_C(\text{core})$ is the correlation energy of the pair of electrons in the $1a_1$ orbital, and is similar to the K-shell correlation energy of $0.041H_\infty$ in the isolated carbon atom. It is generally expected that changes in the correlation energy of atomic inner shells make only a small contribution to, for example, the dissociation energy of a molecule. The intershell correlation energy is only a very small fraction of the total, with an average contribution per spin-orbital pair of $0.0007H_\infty$. On the other hand, the intrapair and interpair contributions in the valence shell are all of comparable magnitudes, in the range $0.005H_\infty$ to $0.01H_\infty$ per spin-orbital pair. This may be understood in terms of the delocalized nature of the $1t_2$ orbitals, and the fact that the $2a_1$ and $1t_2$ orbitals have their maximum densities in the same regions of space. As a result, the total interpair energy is larger than the total intrapair energy. We shall return to this point in §6.4.2, when we shall see that a transformation from delocalized orbitals to localized orbitals leads to an increase of the intrapair energy at the expense of the interpair energy.

74

*The ground state of fluorine* (Das and Wahl 1972*b*). The ground state of the fluorine molecule has the RHF configuration

$$(1\sigma_g)^2(1\sigma_u)^2(2\sigma_g)^2(2\sigma_u)^2(3\sigma_g)^2(1\pi_u)^4(1\pi_g)^4$$

and, as we saw in §3.6, the negative RHF dissociation energy shows that the RHF wave function does not describe a pair of RHF ground-state atoms when the internuclear distance is very large. The reason for this incorrect behaviour is that the $3\sigma_g$ orbital makes a contribution to the spatial part of the wave function which, in simple LCAO terms, is of the form

$$[2p\sigma_A(1)\,2p\sigma_B(2) + 2p\sigma_B(1)\,2p\sigma_A(2)]$$
$$+ [2p\sigma_A(1)\,2p\sigma_A(2) + 2p\sigma_B(1)\,2p\sigma_B(2)]$$

where, for example, $2p\sigma_A$ is a $2p$ orbital on the atom labelled A which points towards the other atom, B. Just as for the $(1\sigma_g)^2$ configuration of $H_2$, this form remains unchanged as the internuclear distance increases and, in the limit of infinite separation of the nuclei, the RHF wave function describes a mixture of states of the systems $(F+F)$ and $(F^+ + F^-)$. This behaviour can be corrected by adding to the RHF configuration a doubly substituted configuration in which the $3\sigma_g$ orbital has been replaced by a $3\sigma_u$ orbital. An MC–SCF calculation (Das and Wahl 1966) involving the two configurations gives a dissociation energy of $+0.020H_\infty$ compared with the RHF value of $-0.050H_\infty$ and the experimental value of $0.062H_\infty$.

Das and Wahl (1972*b*) have analysed the correlation energy of a molecule in terms of three classes of configurations: (1) those necessary for the proper dissociation of the molecule into Hartree–Fock atoms, (2) those that exist in the molecule but vanish formally in the separated atoms, and (3) those that describe mainly atomic correlation and whose contributions are relatively insensitive to changes of internuclear distance. The wave functions considered by Das and Wahl are of the pair-correlated

TABLE 4.4 *Spectroscopic constants for the ground state of fluorine* *(Das and Wahl 1972b)*

| Wave function | $R_e/a_\infty$ | $D_e/H_\infty$ | $\omega_e/\mathrm{cm}^{-1}$ |
|---|---|---|---|
| RHF | 2.50 | −0.050 | 1257 |
| 2-configuration | 2.74 | 0.020 | 678 |
| 6-configuration | 2.64 | 0.067 | 986 |
| 'accurate' | 2.67 | 0.061 | 942 |
| Experiment | 2.68 | 0.062 | 932 |

# Beyond the orbital approximation

form (4.16) but with the orbitals recalculated in the main by the MC–SCF method. The results are summarized in table 4.4. The six-configuration wave function includes four configurations belonging to class (2), whilst the most accurate wave function contains configurations belonging to all three classes. The correlation energy originating from the $2s$ orbitals is found to be very insensitive to changes of internuclear distance, and therefore makes no significant contribution to the dissociation energy and to the potential curve as a whole. The potential curves calculated from the six-configuration and 'accurate' wave functions are closely parallel to the experimental curve; the former is accurate to within $\pm 0.01 H_\infty$, the latter to within $\pm 0.002 H_\infty$.

## 4.5 RELATIVISTIC CORRECTIONS

The Hamiltonian (1.2), or (1.4) in the Born–Oppenheimer approximation, describes a system of non-relativistic structureless particles interacting through electrostatic forces only. Some important physical effects however, notably those met in magnetic resonance spectroscopy, require the explicit consideration of those relativistic properties of the system that arise from the spins of the electrons and nuclei. These are the familiar spin-orbit and spin-spin interactions between the particles, which are often described by an effective 'spin Hamiltonian', and are then treated by perturbation theory as small corrections to the spinless (non-relativistic) Hamiltonian (McWeeny and Sutcliffe 1969, chapter 8; Moss 1973). Other corrections to the non-relativistic model include the magnetic orbit–orbit interactions of the electrons, the dependence of the electronic mass on velocity, and a relativistic correction to the electrostatic potential energy of the electrons. These can be calculated either by perturbation theory (for example, Hartmann and Clementi 1964) or from relativistic (Dirac) Hartree–Fock theory (Maly and Hussonois (1973) have performed such calculations for the ground states of all the elements from $Z = 1$ to $Z = 120$).

The value of a property $P$ of an atom or molecule can be written as

$$P = P_{\mathrm{RHF}} + P_{\mathrm{C}} + P_{\mathrm{R}}$$

where $P_{\mathrm{RHF}}$ is the value in (non-relativistic) Hartree–Fock theory, $P_{\mathrm{C}}$ is the correction due to electron correlation, and $P_{\mathrm{R}}$ is the relativistic correction (terms arising from, for example, the finite sizes of the nuclei should also be included). The quantity $P_{\mathrm{RHF}} + P_{\mathrm{C}}$ is obtained from

TABLE 4.5 *Restricted Hartree–Fock energies* $(E_{\mathrm{RHF}})$, *electron correlation energies* $(E_{\mathrm{C}})$, *and relativistic energies* $(E_{\mathrm{R}})$ *of the ground states of atoms* (*Veillard and Clementi 1968*; *Mann and Johnson 1971*)

| Z | Atom | $-E_{\mathrm{RHF}}/H_\infty$ | $-E_{\mathrm{C}}/H_\infty$ | $-E_{\mathrm{R}}/H_\infty$ |
|---|------|------|------|------|
| 2 | He | 2.8617 | 0.0420 | 0.0001 |
| 3 | Li | 7.4327 | 0.0454 | 0.0006 |
| 4 | Be | 14.5730 | 0.0940 | 0.0022 |
| 5 | B | 24.5291 | 0.1240 | 0.0060 |
| 6 | C | 37.6886 | 0.1551 | 0.0138 |
| 7 | N | 54.4009 | 0.1861 | 0.0273 |
| 8 | O | 74.8094 | 0.2539 | 0.0494 |
| 9 | F | 99.4093 | 0.3160 | 0.0829 |
| 10 | Ne | 128.547 | 0.381 | 0.131 |
| 11 | Na | 161.859 | 0.386 | 0.200 |
| 12 | Mg | 199.615 | 0.428 | 0.295 |
| 13 | Al | 241.877 | 0.459 | 0.421 |
| 14 | Si | 288.854 | 0.494 | 0.584 |
| 15 | P | 340.719 | 0.521 | 0.791 |
| 16 | S | 397.505 | 0.595 | 1.051 |
| 17 | Cl | 459.482 | 0.667 | 1.372 |
| 18 | Ar | 526.817 | 0.732 | 1.761 |
| 30 | Zn | 1777.9 | | 16.8 |
| 36 | Kr | 2752.1 | | 36.8 |
| 82 | Pb | 19524 | < 10† | 1390 |

† A value of $-0.05H_\infty$ per electron gives $E_{\mathrm{C}} = -4.1H_\infty$ for Pb.

the non-relativistic Schrödinger equation and, unless identically zero, its magnitude is usually very large compared with that of $P_{\mathrm{R}}$. Properties that are dominated by the non-relativistic contribution include total and dissociation energies, dipole moments, and vibrational frequencies. The values of atomic electronic energies in table 4.5 show however that the magnitude of the relativistic correction can become large in absolute terms, particularly in systems with large nuclear charges. Thus, whereas the correlation energy increases only slowly with increasing atomic number, and is never greater than about $-10H_\infty$ (for $Z < 100$), the relativistic energy is negligibly small for He, is already larger than the correlation energy for Si, and increases rapidly thereafter: $-1533\,\mathrm{kJ\,mol^{-1}}$ for Si, $-4620\,\mathrm{kJ\,mol^{-1}}$ for Ar, $-9.7 \times 10^4\,\mathrm{kJ\,mol^{-1}}$ for Kr, and $-3.7 \times 10^6\,\mathrm{kJ\,mol^{-1}}$ for Pb. It is clear therefore that some justification of the use of the non-relativistic equation is required for all but the lightest elements.

The contributions to the total relativistic energy of the different

# Beyond the orbital approximation

TABLE 4.6  *Relativistic energies* $(-E_R/H_\infty)$ *of the subshells of atoms*
(*Clementi 1965a*)

| Atom | 1s | 2s | 2p | 3s | 3p | 4s | 3d |
|------|------|------|------|------|------|------|------|
| He | 0.0001 | | | | | | |
| Be | 0.0021 | 0.0002 | | | | | |
| Ne | 0.1066 | 0.0138 | 0.0107 | | | | |
| Mg | 0.2281 | 0.0344 | 0.0313 | 0.0014 | | | |
| Ar | 1.2206 | 0.2353 | 0.2574 | 0.0253 | 0.0224 | | |
| Ca | 1.8846 | 0.3810 | 0.4287 | 0.0512 | 0.0513 | 0.0029 | |
| Zn | 9.9007 | 2.2848 | 2.8023 | 0.3684 | 0.4472 | 0.0151 | 0.1004 |

subshells of electrons in atoms are shown in table 4.6. We see that the bulk of the relativistic energy comes from the inner-shell electrons. This is due to the large mass correction for these electrons, whose speeds approach the speed of light as the nuclear charge increases, and to the strong electric field near the nucleus, which leads to a large correction to the electrostatic potential energy of the inner-shell electrons. All other relativistic contributions are small in comparison to these. In particular, the average relativistic energy per valence electron in any atom is less than $-0.01H_\infty(-26\,\mathrm{kJ\,mol^{-1}})$; for example, $-0.003H_\infty$ in Ne, $-0.006H_\infty$ in Ar, and $-0.008H_\infty$ in Zn, as compared with $-0.05H_\infty$ for a 1s electron in Ne, $-0.6H_\infty$ in Ar, and $-5H_\infty$ in Zn. The accurate computation of, for example, the binding energies of inner-shell electrons therefore requires a proper consideration of the relativistic correction as well as of the correlation energy. In fact, since the magnitude of the correlation correction to an electron binding energy is always less than about $0.03H_\infty$, the relativistic correction is the larger for Ne and subsequent atoms. On the other hand, the relativistic correction (and to a lesser extent the correlation correction) to an inner-shell binding energy is almost independent of the environment of the atom, so that the chemically interesting *differences* of inner-shell binding energies in different molecules can usually be interpreted in terms of the non-relativistic theory (see §6.6).

Many of the properties of molecules that are of interest in chemistry are associated with the valence electrons, and the relativistic corrections are then expected to be of little importance. Thus the relativistic correction to the dissociation energy of $F_2$ is less than $0.001H_\infty$ (see table 4.4), whilst the contribution to the dissociation energy of NaCl has been estimated as $0.002H_\infty$ ($5\,\mathrm{kJ\,mol^{-1}}$), compared with the experimental value

78

of $D_e = 0.16H_\infty$ and the correlation contribution of $0.04H_\infty$ (Matcha 1968). In addition, a comparative non-relativistic (approximate RHF) study of CO and PbO has suggested that the relativistic corrections to molecular properties such as the equilibrium geometry, the dissociation energy and vibration constants are no larger in PbO than in CO (Schwenzer *et al.* 1973). These examples, whilst far from conclusive, suggest that the non-relativistic Schrödinger equation provides a satisfactory model for the computation and interpretation of many molecular properties of interest in chemistry, even when the heavier elements are involved. A final conclusion must however await a more thorough and direct investigation.

# 5 Representation of the orbitals

We have been concerned so far chiefly with the general theoretical tools that are available for the representation and calculation of atomic and molecular wave functions. These wave functions are nearly always expressed in terms of orbitals, either as single configurations in the orbital approximation or, more generally, as linear combinations of configurations. We turn now to the practical problems of the representation and computation of the orbitals. In this chapter we will be concerned principally with the practical implementation of Hartree–Fock theory, and we will consider in particular the solution of the Hartree–Fock equations within the restricted scheme. The generalization for other Hartree–Fock schemes and for multiconfigurational wave functions is straightforward.

## 5.1 MATRIX FORMULATION OF HARTREE–FOCK THEORY

The restricted Hartree–Fock wave function for a closed-shell state of an $N$-electron system is a single Slater determinant made up of $N/2$ doubly occupied orbitals; as in § 3.3.1,

$$\Psi = |\psi_1\alpha, \psi_1\beta, \psi_2\alpha, \psi_2\beta, ..., \psi_{N/2}\alpha, \psi_{N/2}\beta| \qquad (5.1)$$

and the corresponding energy is

$$E = 2\sum_{n=1}^{N/2} f_n + \sum_{m,n=1}^{N/2} (2J_{mn} - K_{mn}) + \sum_{\alpha<\beta=1}^{\nu}\sum \frac{Z_\alpha Z_\beta}{R_{\alpha\beta}} \qquad (5.2)$$

The minimization of the energy with respect to arbitrary variations of the orbitals, subject to the condition that the orbitals be orthonormal, then leads to the RHF equations (3.26)

$$h^{\mathrm{RHF}}\psi_n = \epsilon_n\psi_n \qquad (5.3)$$

*Arbitrary* variations of the orbitals, however, imply either that the orbitals are calculated in numerical form or that the orbitals are analytic functions containing an infinite number of variational parameters. In practice, although accurate numerical solutions have been obtained for atoms (Hartree 1957; Froese 1966), the orbitals do not have such a completely general form, and the minimization of the energy then leads to a set of approximate Hartree–Fock equations.

The most common representation of atomic and molecular orbitals is as linear combinations of some set of known one-electron functions, and from such a set of $M$ linearly independent functions, $\chi_1, \chi_2, ..., \chi_M$, it is possible to construct $M$ linearly independent orbitals

$$\psi_n(\mathbf{r}) = \sum_{i=1}^{M} \chi_i(\mathbf{r})\, C_{in} \quad (n = 1, 2, ..., M) \tag{5.4}$$

The energy (5.2) of a closed-shell state of an $N$-electron system is then a function of the expansion coefficients $C_{in}$ of the $N/2$ occupied orbitals, and the minimization of the energy with respect to these coefficients gives a set of $M$ secular equations of the form

$$\sum_{j=1}^{M} (H_{ij} - \epsilon S_{ij})\, C_j = 0 \quad (i = 1, 2, ..., M) \tag{5.5}$$

where $\qquad H_{ij} = \int \chi_i^* \hbar^{\mathrm{RHF}} \chi_j \, dv, \quad S_{ij} = \int \chi_i^* \chi_j \, dv$

and $\hbar^{\mathrm{RHF}}$ has exactly the same form as the RHF operator

$$h^{\mathrm{RHF}} = f + 2\mathscr{J} - \mathscr{K}$$

but with the $N/2$ occupied orbitals in the Coulomb and exchange operators replaced by the corresponding linear combinations (5.4). Then

$$H_{ij} = f_{ij} + \sum_{m=1}^{N/2} \sum_{k=1}^{M} \sum_{l=1}^{M} C_{km}^* C_{lm} \{2[ij|kl] - [il|kj]\} \tag{5.6}$$

where $\qquad f_{ij} = \int \chi_i^* f \chi_j \, dv$

$$[ij|kl] = \int\int \frac{\chi_i^*(1)\,\chi_j(1)\,\chi_k^*(2)\,\chi_l(2)}{r_{12}} \, dv_1 \, dv_2 \tag{5.7}$$

We can now repeat the argument outlined in §1.3. A non-trivial solution of the equations (5.5) is obtained if the orbital energy $\epsilon$ is chosen such that the secular determinant, whose elements are $(H_{ij} - \epsilon S_{ij})$, is zero:

$$\det(H_{ij} - \epsilon S_{ij}) = 0 \tag{5.8}$$

The secular determinant is a polynomial of degree $M$ in the orbital energy, and it has $M$ roots,

$$\epsilon_1 \leqslant \epsilon_2 \leqslant \epsilon_3 \leqslant ... \leqslant \epsilon_M$$

Corresponding to each orbital energy $\epsilon_n$, an orbital (5.4) is obtained by

## Representation of the orbitals

solving the secular equations and normalization. The resulting orbitals are orthogonal. If the state of interest is a closed-shell ground state, the corresponding total wave function (5.1) is then normally obtained by choosing those $N/2$ orbitals with the lowest orbital energies to be doubly occupied.

The secular equations (5.5) can be written in the matrix form

$$\boldsymbol{HC} = \epsilon \boldsymbol{SC} \qquad (5.9)$$

in which $\boldsymbol{H}$ and $\boldsymbol{S}$ are the square $(M \times M)$ matrices whose elements are $H_{ij}$ and $S_{ij}$, and $\boldsymbol{C}$ is a column vector whose elements are the coefficients $C_j$. The matrix $\boldsymbol{H}$ is a representation of the Hartree–Fock operator $h^{\mathrm{RHF}}$ in terms of the *basis* of $M$ functions $\chi_i$, and its eigenvectors give $M$ approximate RHF orbitals. Increasing the number of basis functions leads to better approximations to an increasing number of orbitals. When $M$ becomes infinitely large, and the basis becomes complete, the matrix equation (5.9) becomes entirely equivalent to the Hartree–Fock eigenvalue equation $h^{\mathrm{RHF}}\psi = \epsilon\psi$, having the same set of eigenvalues and eigenfunctions, the latter being expressed in the form (5.4) as linear combinations of the basis functions.

Although we have derived the matrix formulation of the restricted Hartree–Fock scheme for closed-shell states only, the derivation for open-shell states is entirely analogous and introduces no new features.

*The self-consistent field method.* Just as the Hartree–Fock operator $h^{\mathrm{RHF}}$ is a function of the occupied orbitals (§3.2.1), its representative matrix $\boldsymbol{H}$ is a function of the expansion coefficients $C_{in}$ of the occupied orbitals. It is therefore necessary to solve the matrix eigenvalue problem (5.9) by some iterative method and, as discussed in §3.2.1, the method most commonly used is the SCF method, which can be summarized as

$$\ldots \to \boldsymbol{C}_n^{(i)} \to \boldsymbol{H}^{(i)} \to \boldsymbol{C}_n^{(i+1)} \to \boldsymbol{H}^{(i+1)} \to \ldots$$

An initial Hamiltonian matrix $\boldsymbol{H}^{(0)}$ is calculated from some suitably chosen initial set of vectors $\boldsymbol{C}_n^{(0)}$ for the occupied orbitals. The eigenvectors $\boldsymbol{C}_n^{(1)}$ of $\boldsymbol{H}^{(0)}$ are then used to construct a new matrix $\boldsymbol{H}^{(1)}$ with eigenvectors $\boldsymbol{C}_n^{(2)}$. This iterative cycle is continued until the solutions are self-consistent; that is, until the eigenvectors $\boldsymbol{C}_n^{(i+1)}$ and eigenvalues $\epsilon_n^{(i+1)}$ are the same, to the required accuracy, as $\boldsymbol{C}_n^{(i)}$ and $\epsilon_n^{(i)}$.

Because the matrix equation (5.9) cannot, for a finite basis, give exact Hartree–Fock orbitals, the approximate RHF equations which it represents are often distinguished from the formally exact RHF equations (5.3) by some alternative name. The most commonly used names are SCF equations, RHF–SCF equations, or Roothaan equations (in recognition of the author of one of the first general matrix formulations of Hartree–Fock theory; Roothaan 1951, 1960). The solutions are usually called SCF orbitals and energies, the prefix RHF often being reserved for the exact solutions of (5.3).

Given a basis of one-electron functions $\chi_i$, the Hartree–Fock problem is reduced to the evaluation of the matrix elements $H_{ij}$ and $S_{ij}$, and to the iterative solution of the matrix eigenvalue equation (5.9). We are still left however with the problem of the choice of basis functions.

## 5.2 ATOMIC ORBITALS

For the purposes of atomic SCF, and CI, calculations, the atomic orbitals (AO) are most conveniently represented as linear combinations of simple exponential functions called Slater functions or *Slater-type orbitals* (STO) which have the form

$$\chi_{nlm}(\boldsymbol{r}, \zeta) = R_{nl}(r, \zeta) Y_{lm}(\theta, \varphi) \qquad (5.10)$$

where
$$R_{nl}(r, \zeta) = r^{n-1} e^{-\zeta r} \quad (n > l) \qquad (5.11)$$

The angular function $Y_{lm}(\theta, \varphi)$ is a normalized spherical harmonic, familiar from the solutions of the hydrogen-atom problem, and it determines the spatial symmetry properties of the orbital. The STO are given the usual names appropriate to the labels $n$, $l$, $m$; for example, the values $n = 2$, $l = 0$ define a $2s$ Slater-type orbital, and $n = 3$, $l = 1$ define a $3p_m$ Slater-type orbital. The exponent $\zeta$ is a parameter which can be calculated variationally in any particular atomic calculation. An atomic orbital with (orbital) angular momentum quantum numbers $l$ and $m$ is then a linear combination of the basis functions (5.10):

$$\psi = R(r) Y_{lm}(\theta, \varphi) \qquad (5.12)$$

where
$$R(r) = \sum_n \sum_i C_{ni} R_{nl}(r, \zeta_i) \qquad (5.13)$$

the summation being over a set of integer values of $n$ and, for each $n$, a set of values of the exponents $\zeta_i$. Given a set of STO with specified values of the exponents (zeta-values), the expansion coefficients are calculated as solutions of the SCF equations. Optimization of the exponents requires a separate solution of the SCF equations for each set of zeta-values; the optimum set is that which gives the lowest total energy.

Accurate RHF calculations of this type have been performed for the ground states and some low-lying excited states of many atoms and ions (Clementi 1965*b*; Huzinaga 1971). One example is the ground-state calculation for the helium atom discussed in §3.2. Total SCF energies obtained from three sets of calculations of different accuracies for the ground states of the atoms helium to argon are compared in table 5.1. It is found quite generally that a basis of three to five STO per AO is required to produce highly accurate SCF wave functions and energies which are close to the RHF limit. Accuracies that are adequate for many purposes in chemistry, however, can be obtained from a basis of two STO per AO; this is called a *double-zeta* (DZ) basis. A wave function built up from a basis of one STO per AO, a *minimal* basis, is less accurate but often gives a good qualitative representation of the system. An appreciation of the relative accuracies of the minimal and double-zeta bases is of some importance since these bases are often used as measures of the quality of molecular-orbital wave functions. We note that, for the ground

TABLE 5.1 *SCF energies (in units of $H_\infty$) of the ground states of the first- and second-row atoms calculated in terms of minimal (1 STO per AO), double-zeta (2 STO per AO), and accurate (3 to 5 STO per AO) bases*

|  | Minimal (a, b) | Double-zeta (b) | Accurate (b, c) |
|---|---|---|---|
| He | −2.84765 | −2.86167 | −2.86168 |
| Li | −7.41848 | −7.43272 | −7.43273 |
| Be | −14.55674 | −14.57237 | −14.57302 |
| B | −24.49837 | −24.52792 | −24.52906 |
| C | −37.62239 | −37.68675 | −37.68862 |
| N | −54.26890 | −54.39795 | −54.40093 |
| O | −74.54037 | −74.80432 | −74.80930 |
| F | −98.94211 | −99.40130 | −99.40933 |
| Ne | −127.81218 | −128.53508 | −128.54708 |
| Na | −161.12392 | −161.85002 | −161.85889 |
| Mg | −198.85779 | −199.60702 | −199.61461 |
| Al | −241.15376 | −241.87320 | −241.87669 |
| Si | −288.08996 | −288.85120 | −288.85434 |
| P | −339.90988 | −340.71597 | −340.71876 |
| S | −396.62762 | −397.50230 | −397.50487 |
| Cl | −458.52369 | −459.47962 | −459.48204 |
| Ar | −525.76525 | −526.81512 | −526.81748 |

References: (a) Clementi and Riamondi 1963; (b) Huzinaga and Arnau 1970*a*; Huzinaga 1971; (c) Clementi 1965*b*.

states of the atoms up to argon, the absolute error in the SCF energy for the double-zeta basis is on average about $0.004H_\infty$ ($10.5\,\text{kJ mol}^{-1}$) and is never greater than about $0.012H_\infty$ ($31.5\,\text{kJ mol}^{-1}$). For the minimal basis, on the other hand, the error is on average $0.5H_\infty$ ($1313\,\text{kJ mol}^{-1}$), is never less than $0.014H_\infty$ ($36.8\,\text{kJ mol}^{-1}$), and for argon is $1.05H_\infty$ ($2757\,\text{kJ mol}^{-1}$). The energies of interest in chemistry, such as excitation energies and energies of reaction, are seldom greater than $1H_\infty$ ($2626\,\text{kJ mol}^{-1}$), and are often much smaller. They are therefore comparable in magnitude to the errors in the minimal basis SCF energies.

TABLE 5.2  *Orbital exponents $\zeta$ of the optimized minimal and double-zeta STO basis sets for the atoms helium to neon*

| | Minimal (a, b) | | | Double-zeta (b, c) | | | | | |
|---|---|---|---|---|---|---|---|---|---|
| | $1s$ | $2s$ | $2p$ | $1s$ | $1s'$ | $2s$ | $2s'$ | $2p$ | $2p'$ |
| He | 1.6875 | | | 1.4546 | 2.9156 | | | | |
| Li | 2.6906 | 0.6396 | | 2.4823 | 4.6875 | 0.6716 | 1.9757 | | |
| Be | 3.6848 | 0.9560 | | 3.3476 | 5.5430 | 0.5886 | 1.0090 | | |
| B | 4.6794 | 1.2881 | 1.2107 | 4.2448 | 6.5450 | 0.8788 | 1.4142 | 1.0043 | 2.2116 |
| C | 5.6727 | 1.6083 | 1.5679 | 5.1117 | 7.4831 | 1.1635 | 1.8366 | 1.2549 | 2.7238 |
| N | 6.6651 | 1.9237 | 1.9170 | 5.9989 | 8.5276 | 1.4147 | 2.2523 | 1.4961 | 3.2390 |
| O | 7.6579 | 2.2458 | 2.2266 | 6.8758 | 9.5507 | 1.6603 | 2.6709 | 1.6555 | 3.6856 |
| F | 8.6501 | 2.5638 | 2.5500 | 7.7159 | 10.5136 | 1.9333 | 3.1202 | 1.8470 | 4.1746 |
| Ne | 9.6421 | 2.8792 | 2.8792 | 8.6081 | 11.5976 | 2.1619 | 3.5247 | 2.0530 | 4.6784 |

References: (a), (b) and (c) as in table 5.1.

In table 5.2 are shown values of the exponents $\zeta$ of the minimal and double-zeta basis sets obtained from SCF calculations for the ground states of helium and the first-row atoms, lithium to neon. We see that each STO of the minimal basis is replaced in the DZ basis by a pair of functions whose zeta-values (except for the $2s$ function of Li) lie on opposite sides of the minimal basis value. This replacement of the single STO by an 'inner' and an 'outer' function is particularly valuable for the use of the DZ basis in a molecular calculation, since it allows the atomic orbitals to contract or expand in the molecular environment.

### 5.3  THE LCAO–MO METHOD

The most generally applicable and useful representation of a molecular orbital (MO) is as a linear combination of atomic orbitals (LCAO). In the precomputational era of quantum chemistry (before about 1960), the atomic orbitals were often envisaged either as the Hartree–Fock orbitals of the separate atoms or as simple, usually single-STO, representations of them. In current molecular-orbital calculations, an 'atomic orbital' is understood to be any convenient one-centre basis function which is normally, but not invariably, centred on a nucleus of the molecule. Denoting such a basis function by $\chi_i$, the MO have the form

$$\psi_n = \sum_i C_{in} \chi_i$$

and, in the orbital approximation, the coefficients $C_{in}$ are calculated by the SCF procedure. Two types of basis, each with its own advantages and disadvantages, are used for the construction of molecular orbitals: Slater functions and Gaussian functions.

## 5.4 SLATER-TYPE MOLECULAR ORBITALS

For molecular orbitals built up from STO, the 'atomic orbitals' can be of several types:

(i)   the SCF orbitals calculated for the separate atoms;

(ii)  the individual STO obtained from the atomic calculations;

(iii) the STO of (ii) but with the exponents reoptimized for the molecule under consideration;

(iv)  the STO of (ii) or (iii) supplemented by additional functions of lower symmetries to describe the distortion or polarization of the atomic orbitals in the molecular environment; for example, $p$-type functions to describe the polarization of $s$ orbitals, $d$-type functions for the polarization of $p$ orbitals.

Accuracy and flexibility increases from type (i) to type (iv), as does the computational labour.

Results from several SCF calculations using different STO bases are compared in table 5.3 for the $^1\Sigma^+$ closed-shell ground state of carbon monoxide. A number of general conclusions may be drawn from these. The first is that the reoptimization, in the molecule, of the exponents $\zeta$ of the minimal basis does not normally lead to a significant improvement of the energy or other properties of the molecule. This is even more true when double-zeta and larger basis sets are used. For this reason, and the considerable computational effort that would otherwise be involved, molecular orbitals are normally constructed from the atomic basis functions with the atomic values of $\zeta$. The only important exception in general is the $1s$ orbital of the hydrogen atom, whose exponent is in-

TABLE 5.3 *Total energy, dissociation energy and dipole moment for several SCF and CI calculations in terms of STO for the ground state of carbon monoxide at the experimental internuclear distance of $2.132a_\infty$*

| Method | Reference | $-E/H_\infty$ | $D_e/H_\infty$ | $\mu/D$ |
|---|---|---|---|---|
| SCF | | | | |
| Minimal with atomic $\zeta$ | a | 112.326 | 0.163 | −0.59 |
| Minimal with optimized $\zeta$ | b | 112.391 | 0.228 | −0.46 |
| DZ with atomic $\zeta$ | c | 112.676 | 0.185 | +0.60 |
| DZ with atomic $\zeta$ + one set of $3d$ polarization functions, | c | 112.770 | 0.279 | +0.14 |
| Accurate SCF (RHF) | c | 112.789 | 0.291 | +0.28 |
| | | | | |
| CI† | | | | |
| SCF + 200 $D$ | d | 113.0338 | | +0.18 |
| SCF + 138 $D$ + 62 $S$ | d | 113.0179 | | −0.08 |
| SCF + 117 $D$ + 36 $S$ | e | 113.0151 | | −0.12 |
| | | | | |
| Experimental (C⁻O⁺) | | 113.377 | 0.413 | −0.11 |

† $D$ stands for a doubly substituted configuration and $S$ for a single substituted configuration.

References: (*a*) Ransil 1960; (*b*) Huo 1965; (*c*) Yoshimine and McLean 1967; (*d*) Grimaldi, Lecourt and Moser 1967; (*e*) Green 1971.

creased from its value of 1.0 in the isolated atom to about 1.2 in the molecule; this increase describes a contraction of the orbital in the molecular environment, and has been found to be important for the correct description of the charge distribution near a proton in a molecule. The second general conclusion is that values of properties comparable with accurate RHF values cannot normally be obtained with a basis smaller than a double-zeta basis plus some polarization functions. For the atoms B to Ne, such a basis consists of the atomic DZ set plus a set of five $3d$ functions to describe the polarization of the $2p$ atomic orbitals in the molecular environment. For the hydrogen atom, the corresponding basis is made up of two $s$-type STO and a set of $2p$ polarization functions.

We have already noted in §3.5.1 that, although the error in the RHF value of the dipole moment of CO is characteristic of the error expected for nonpolar diatomic molecules, the moment has the wrong polarity (the correct sign obtained with the minimal basis is a fortuitous result). Results from several CI calculations, all yielding about 40 per cent of the correlation energy, are shown in table 5.3. It is clear that accuracy of the total energy does not in general represent a valid criterion for the accuracies of other properties. Most interesting is the result that although the addition of doubly substituted configurations gives a significant lowering of the energy, it does not lead to a significant change of the dipole moment from its RHF value. The correct magnitude and sign is obtained only when singly substituted terms are added. Thus, whereas Brillouin's theorem for closed-shell states implies that the most significant contributions to the energy come from the doubly substituted configurations, it is a general conclusion that the singly substituted configurations, which enter the wave function only by coupling with the doubly substituted configurations, cause the major changes in one-electron properties such as the dipole moment. These changes are nearly always small however. For example, the dipole moment of CO is the difference between two large contributions from the nuclei and the electrons, and the change in the total value from $+0.28D$ to $-0.11D$ represents only a small fractional change of the electronic contribution. This demonstrates again the fact that the orbital approximation can be highly accurate except for those properties whose values are essentially differences between pairs of much larger quantities.

## 5.5 GAUSSIAN-TYPE MOLECULAR ORBITALS

Although it has been demonstrated that Slater functions form a satisfactory basis for the representation of molecular orbitals, accurate SCF and CI calculations in terms of them have been performed only for diatomic and small polyatomic molecules, with few non-empirical calculations of any kind for molecules larger than methane. The reason for this apparent interest in small molecules is the great difficulty associated with the evaluation of the electron-interaction integrals (5.7)

$$[ij|kl] = \int\int \frac{\chi_i^*(1)\chi_j(1)\chi_k^*(2)\chi_l(2)}{r_{12}} \, dv_1 \, dv_2$$

which occur, for example, in the matrix elements of the SCF Hamiltonian matrix $H$ (§5.1). The difficulties are most severe for three- and four-centre integrals when the Slater functions are centred on three or four non-collinear atoms in a molecule. The existence of this 'integral bottleneck' has led to an increasing use of the Gaussian-type orbitals (GTO) introduced into quantum chemistry by Boys (1950) as an alternative to STO. Wave functions for polyatomic molecules are now almost exclusively treated in this way.

Gaussian-type orbitals are one-centre functions of the type (5.10) but with radial factors
$$R_{nl}(r, \gamma) = r^{n-1} e^{-\gamma r^2} \quad (n > l) \tag{5.14}$$

When the number $(n - l - 1)$ is restricted to zero or even-integer values, these functions have a very clear advantage over Slater functions with regard to the evaluation of electron-interaction integrals, since the product of two Gaussians on different centres is equivalent to a single Gaussian on a new centre. An integral like (5.7) in which the four $\chi$s are Gaussians on different centres then reduces immediately to a two-centre integral which can be evaluated explicitly in terms of known functions.

*Example. A four-centre integral.* We consider an electron-interaction integral (5.7) in which the atomic orbitals are $1s$ Gaussians on different centres:
$$[aA, bB|cC, dD] = \int\int \frac{\chi(r_{A1}, a)\, \chi(r_{B1}, b)\, \chi(r_{C2}, c)\, \chi(r_{D2}, d)}{r_{12}}\, dv_1\, dv_2$$
where, for example,
$$\chi(r_{A1}, a) = \exp(-ar_{A1}^2)$$

is a $1s$ Gaussian (unnormalized) centred on the point $A$ in fig. 5.1. Let $\mathbf{R}_A$, $\mathbf{R}_B$ be the position vectors of points $A$, $B$ relative to the origin of a coordinate system, and let the point $P$ be defined by
$$(a+b)\mathbf{R}_P = a\mathbf{R}_A + b\mathbf{R}_B$$
Then the product of the functions centred on $A$ and $B$ is
$$\exp(-ar_{A1}^2 - br_{B1}^2) = \exp\left(\frac{-ab}{a+b} R_{AB}^2\right) \chi(r_{P1}, a+b)$$

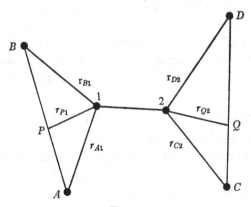

Fig. 5.1

*Representation of the orbitals*

where
$$\chi(r_{P1}, a+b) = \exp\left[-(a+b)\,r_{P1}^2\right]$$
is a $1s$ Gaussian on the new centre $P$. Similarly,

$$\chi(r_{C2}, c)\,\chi(r_{D2}, d) = \exp\left(\frac{-cd}{c+d}R_{CD}^2\right)\chi(r_{Q2}, c+d)$$

where $(c+d)\,R_Q = c R_C + d R_D$. Then

$$[aA, bB \,|\, cC, dD] = \exp\left(\frac{-ab}{a+b}R_{AB}^2 - \frac{cd}{c+d}R_{CD}^2\right)$$
$$\times \iint \frac{\chi(r_{P1}, a+b)\,\chi(r_{Q2}, c+d)}{r_{12}}\,dv_1\,dv_2$$

and the four-centre integral has been reduced to a two-centre integral. This may be further reduced by standard methods (Boys 1950; Shavitt 1963) to give

$$[aA, bB \,|\, cC, dD] = \frac{2\pi^{\frac{5}{2}}}{(a+b)(c+d)(a+b+c+d)^{\frac{1}{2}}}$$
$$\times F\left[\frac{(a+b)(c+d)}{(a+b+c+d)}R_{PQ}^2\right]\exp\left(\frac{-ab}{a+b}R_{AB}^2 - \frac{cd}{c+d}R_{CD}^2\right) \quad (5.15)$$

where
$$F[z] = \frac{1}{\sqrt{z}}\int_0^{\sqrt{z}} e^{-x^2}\,dx = \frac{1}{2}\left(\sqrt{\frac{\pi}{z}}\right)\operatorname{erf}(\sqrt{z})$$

and $\operatorname{erf}(\sqrt{z})$ is the well-known error function (Abramowitz and Stegun 1965).

When the angular dependence of an atomic orbital is described in the usual way by a spherical harmonic $Y_{lm}$, the corresponding molecular integrals are similar to, but a little more complicated than, (5.15). This added complication may be avoided by expressing an atomic or molecular orbital in terms of the simple spherical functions $e^{-\gamma r^2}$ only. The angular dependence of a GTO with $l > 0$ is then simulated by the superposition of a number of symmetrically distributed spherical Gaussians. Thus a $p_z$ orbital is represented by the superposition of two functions, each representing a lobe:

$$p_z(r) = \exp\left[-\gamma(x^2+y^2+(z-R)^2)\right] - \exp\left[-\gamma(x^2+y^2+(z+R)^2)\right]$$

where, if the centre of the orbital is taken as the origin, the Gaussians are centred at positions $(0, 0, \pm R)$. A $p$-type GTO is therefore specified by the two parameters $\gamma$ and $R$. Similarly, a $d$-type orbital is represented as a superposition of four spherical Gaussians. Such *Gaussian lobe functions* (Whitten 1963) are then used in exactly the same way as the more

90

conventional GTO, and the two types are highly comparable with respect both to computational labour and to accuracy (Shih *et al.* 1970).

The main disadvantage in using a Gaussian basis is that about two to five times as many Gaussian functions as Slater functions are required to produce a given accuracy. In table 5.4 are shown the results of orbital calculations for the ground states of H, He and Be, in which the *s* orbitals have been expressed as linear combinations of 1*s* Gaussians:

$$ns \simeq \sum_{i=1}^{N} C_i e^{-\gamma_i r^2}$$

We see that ten Gaussians are required to produce six-figure accuracy for the energy of the hydrogen atom and double-zeta accuracy for He and Be (table 5.1). More generally, a basis of ten 1*s*-type and six sets of 2*p*-type Gaussians is required to produce double-zeta accuracy for the first-row atoms. A basis of twelve 1*s*-type and nine sets of 2*p*-type Gaussians for the second-row atoms Al to Ar is almost as good as a double-zeta Slater-function basis. The main reason for this increase in the basis size on going from Slater to Gaussian functions is the incorrect behaviour of the latter at and near a nucleus (Steiner and Sykes 1972). A Slater function like $e^{-\zeta r}$ has a non-zero derivative at $r = 0$,

$$\left[ \frac{d}{dr} e^{-\zeta r} \right]_{r=0} = -\zeta \neq 0$$

and this *cusp* property is essential for the accurate description of an atomic or molecular wave function at and near a nucleus. The

TABLE 5.4  *SCF energies (in units of $H_\infty$) of the ground states of the hydrogen, helium and beryllium atoms, calculated with a basis of N 1s-type Gaussians*

| N | H (a) | He (a) | Be (b) |
|---|---|---|---|
| 1 | −0.424413 | −2.3009869 | |
| 2 | −0.485813 | −2.7470661 | |
| 3 | −0.496979 | −2.8356798 | |
| 4 | −0.499277 | −2.8551603 | |
| 5 | −0.499809 | −2.8598949 | |
| 6 | −0.499940 | −2.8611163 | −14.556374 |
| 7 | −0.499976 | −2.8614912 | −14.567118 |
| 8 | −0.499991 | −2.8616094 | −14.571161 |
| 9 | −0.499997 | −2.8616523 | −14.572080 |
| 10 | −0.499999 | −2.8616692 | −14.572580 |
| 11 | | | −14.572842 |

References: (*a*) Huzinaga 1965; (*b*) Huzinaga 1971.

corresponding Gaussian $e^{-\gamma r^2}$ has zero derivative at $r = 0$. Of the ten Gaussians required to describe the $1s$ and $2s$ orbitals of the first-row atoms Li to Ne, five or six are used for the region very close to the nucleus, and only about two Gaussians per orbital are required for the intermediate to large distances from the nucleus. Similarly, for the $2p$ orbital in these atoms, about three $p$-type Gaussians are used for the region close to the nucleus and three for the intermediate to large distances from the nucleus. Apart from the region close to a nucleus therefore, the Gaussian basis is almost as good as the Slater-function basis.

### 5.5.1 Contracted Gaussians

We saw in §5.4 that the double-zeta Slater-function basis calculated for an atom forms a satisfactory basis for the representation of the atom in a molecule, particularly when polarization functions are included to describe the polarization of the atomic orbitals in the molecular environment. Gaussian basis sets of various accuracies, including double-zeta accuracy, have been calculated for many atoms by the SCF method (Huzinaga 1965, 1971; Veillard 1968; Whitman and Hornback 1969; Roos and Seigbahn 1970; Wachters 1970). These can be supplemented by Gaussian polarization functions (Dunning 1971a) and used for the construction of molecular orbitals in the same way as the Slater-function sets. In particular, a Gaussian basis of double-zeta accuracy plus polarization functions can be expected to give accuracies for molecules close to the RHF limit. On the other hand, a Gaussian basis set of double-zeta accuracy for the first-row atoms B to Ne consists of ten $s$-type functions and six sets of $p$-type functions, and the total number $N_G$ of Gaussians in this basis set is therefore two to three times as large as the number $N_S$ of Slater functions in the double-zeta set. Because the number of electron-interaction integrals (5.7) is proportional to the fourth power of the basis size, a ratio of, say, $N_G/N_S = 3$ increases the number of these integrals by a factor of 81. The problem of evaluating a relatively small number of difficult integrals over Slater functions has therefore been replaced by that of evaluating and manipulating a much larger number of simple integrals over Gaussian functions. A calculation for a molecule of quite modest dimensions can involve several hundred Gaussians and $10^8$ to $10^{10}$ integrals.

One way of reducing this considerable computational problem to manageable proportions is by the use of *contracted* Gaussian functions. A

TABLE 5·5  *The ground state of the water molecule. Total and dissociation energies, dipole moment and geometry for several basis sets in the RHF approximation*

| Basis | Reference | $-E/H_\infty$ | $D_e/H_\infty$ | $\mu/D$ | $R_{OH}/a_\infty$† | $\angle HOH$† |
|---|---|---|---|---|---|---|
| **SLATER** | | | | | | |
| 1 Minimal : O(2s, 1p), H(1s) | a | 75.7055 | 0.165 | | 1.871 | 100.3° |
| 2 O(5s, 3p), H(2s) | b | 76.0182 | | 2.71 | | |
| 3 O(5s, 3p, 1d), H(2s, 1p) | b | 76.0593 | | 2.09 | | |
| 4 O(5s, 4p, 1d), H(3s, 1p) | b | 76.0631 | 0.254 | 2.05 | | |
| **GAUSSIAN** | | | | | | |
| 5 Uncontracted: O(9s, 5p), H(4s) | c | 76.0133 | 0.215 | 2.65 | | |
| 6 DZ: O(9s, 5p/4s, 2p), H(4s/2s) | c | 76.0093 | | 2.68 | | |
| 7 O(9s, 5p, 2d/4s, 3p, 2d), H(4s, 1p/2s, 1p) | b | 76.0510 | 0.254 | | 1.779 | 106.1° |
| 8 O(11s, 7p/6s, 5p), H(5s/3s) | b | 76.0242 | | 2.72 | | |
| 9 O(11s, 7p, 2d/6s, 5p, 2d), H(5s, 1p/3s, 1p) | b | 76.0628 | 0.255 | 2.03 | 1.779 | 106.6° |
| **EXPERIMENT** | | | 0.370 | 1.85 | 1.809 | 104.5° |

† Only where the bond distance and angle are shown has the geometry been optimized. In all other cases the assumed geometry was $R_{OH} = 1.811a_\infty$, $\angle$ HOH = 104.45°.

References: (a) Pitzer and Merrifield 1970; (b) Dunning, Pitzer and Aung 1972; (c) Dunning 1970a.

## Representation of the orbitals

contracted Gaussian is a linear combination of (*primitive*) Gaussians with fixed coefficients (Whitten 1966; Petke, Whitten and Douglas 1969; Salez and Veillard 1968; Huzinaga and Arnau 1970*b*; Dunning 1970*a, b,* 1971*b*). For example, the ten *s*-type Gaussians of the double-zeta set for the first-row atoms can be contracted into four linear combinations which simulate the behaviour of the four Slater functions used for the description of the 1*s* and 2*s* orbitals in these atoms. Similarly, the six sets of *p*-type Gaussians can be contracted into two sets of linear combinations to simulate the Slater double-zeta description of the 2*p* atomic orbital. Such a contraction is denoted by (10*s*, 6*p*/4*s*, 2*p*). The use of fixed coefficients in the contracted functions necessarily leads to some loss of accuracy in a molecular calculation, but the results are nevertheless normally almost as close to the Hartree–Fock limit as those obtained with the Slater double-zeta basis.

The results of SCF calculations for the ground state of the water molecule are shown in table 5.5 for nine different basis sets. The effect of contraction, as illustrated by the sets 5 and 6, is seen to be very small, both for the total energy and the dipole moment. The exponents of the primitive Gaussians and the contraction coefficients of the Gaussian set (9*s*, 5*p*/4*s*, 2*p*) of the oxygen atom and the set (4*s*/2*s*) of the hydrogen atom

TABLE 5.6 *Gaussian exponents $\gamma_i$ and contraction coefficients $C_i$ for the basis* O(9*s*, 5*p*/4*s*, 2*p*), H(4*s*/2*s*). *The Gaussians are of types* 1*s* *and* 2*p* *only. Each contracted function is a normalized combination of normalized primitive Gaussians* (Dunning 1970*a*)

| Oxygen *s* set | | Oxygen *p* set | | Hydrogen *s* set | |
|---|---|---|---|---|---|
| $\gamma_i$ | $C_i$ | $\gamma_i$ | $C_i$ | $\gamma_i$ | $C_i$ |
| 7816.5400 | 0.002031 | 35.1832 | 0.019580 | 19.2406 | 0.032828 |
| 1175.8200 | 0.015436 | 7.9040 | 0.124189 | 2.8992 | 0.231208 |
| 273.1880 | 0.073771 | 2.3051 | 0.394727 | 0.6534 | 0.817238 |
| 81.1696 | 0.247606 | 0.7171 | 0.627375 | | |
| 27.1836 | 0.611832 | | | 0.1776 | 1 |
| 3.4136 | 0.241205 | 0.2137 | 1 | | |
| | | | | | |
| 9.5322 | 1 | | | | |
| | | | | | |
| 0.9398 | 1 | | | | |
| | | | | | |
| 0.2846 | 1 | | | | |

are shown in table 5.6. Of the nine primitive $1s$ Gaussians, six are combined to form a contracted function which simulates the behaviour of the innermost Slater function of the DZ set of the oxygen atom, whilst the remaining three functions are left uncontracted. Similarly, four of the five $2p$ Gaussians are combined. This has the dual effect of providing, at the DZ level, an accurate description of the wave function in the region near the oxygen nucleus whilst allowing maximum variational freedom in the bonding region. The results shown in table 5.5 also emphasize the importance of including polarization functions. Not only do these provide a substantial lowering of the total energy, but a comparison of the values of the dipole moment for the sets 6, 8 and 9 shows that the polarization functions are essential for the accurate (RHF) description of such one-electron properties. The computation of a number of molecular properties with several different basis sets is discussed in the following sections.

## 5.6 THE MINIMAL BASIS

We have seen that values of molecular properties close to the RHF values cannot normally be obtained with a basis smaller than a double-zeta basis plus polarization functions. It has been found, however, that molecular-orbital calculations in terms of the minimal basis (one STO or contracted GTO for each occupied atomic orbital) are often capable of providing a good qualitative description of the trends and relative values of properties amongst a series of related systems, and that they can be highly accurate for some molecular properties, such as the ground-state equilibrium geometry.

SCF calculations in terms of a minimal basis of Slater functions have been performed for a number of polyatomic molecules, notably the boron hydrides and related systems by Lipscomb and co-workers (for example, $B_2H_6$: Law, Stevens and Lipscomb 1972; $B_4H_{10}$, $B_5H_9$ and $B_5H_{11}$: Switkes *et al.* 1970a, b; $B_8H_{12}$, $B_9H_{15}$, $B_6H_6^{2-}$, $B_{10}H_{10}^{2-}$ and $B_{10}H_{14}^{2-}$: Hall, Marynick and Lipscomb 1974). The great majority of recent calculations have however been performed in terms of Gaussian functions, the basis functions for a molecular calculation being contracted Gaussians; that is, linear combinations of (primitive) Gaussians designed to simulate the behaviour of the Slater functions of the minimal basis. These linear combinations are usually constructed in one of two ways. One way, as discussed in § 5.5.1, is to start with sets of Gaussians obtained from atomic

## Representation of the orbitals

SCF calculations, and to contract these sets to minimal basis size, either by inspection or by trial calculations on small molecules. A basis of this type has been used, for example, by Clementi and co-workers in a series of SCF calculations on $\pi$-electron systems such as pyrrole and pyridine (Clementi 1968), and adenine, cytosine, guanine and thymine (Clementi *et al.* 1969). The basis used in this series is made up of fixed sets of contracted Gaussians: $(7s, 3p/2s, 1p)$ on C, N and O, and $(3s/1s)$ on H. The use of such a consistent basis is particularly valuable in allowing meaningful comparisons to be made of the properties of related systems.

A somewhat different approach to the construction of the molecular basis is to simulate a Slater-function calculation directly by representing each Slater function of the minimal basis as a fixed linear combination of a given small number of Gaussian functions. These Gaussian representations of Slater functions are determined most directly by a least-squares fitting procedure, and this procedure has been extensively investigated particularly by Pople and co-workers (for example: Hehre, Stewart and Pople 1969; Stewart 1970; Lathan *et al.* 1974). The results of several minimal basis SCF calculations of this type are shown in table 5.7 (Hehre *et al.* 1969). A basis in which each Slater function is represented

TABLE 5.7 *Minimal basis SCF calculations in which each STO is a linear combination of $N$ ($= 2$ to $6$) Gaussians. The results on the extreme right are the exact values for the minimal Slater function sets (Hehre, Stewart and Pople 1969)*

| Molecule | STO–2G | STO–3G | STO–4G | STO–5G | STO–6G | STO |
|---|---|---|---|---|---|---|
| | | | SCF total energies (Hartrees) | | | |
| HF | −95.5572 | −98.5274 | −99.2204 | −99.3969 | −99.4501 | −99.4785 |
| NH₃ | −53.8350 | −55.4536 | −55.8490 | −55.9553 | −55.9874 | −56.0050 |
| H₂CO | −109.0037 | −112.3295 | −113.1362 | −113.3497 | −113.4146 | −113.4496 |
| | | | SCF atomization energies (Hartrees) | | | |
| HF | 0.0683 | 0.0492 | 0.0476 | 0.0469 | 0.0469 | 0.0469 |
| NH₃ | 0.3482 | 0.3085 | 0.2999 | 0.3003 | 0.3000 | 0.2999 |
| H₂CO | 0.4129 | 0.3491 | 0.3383 | 0.3387 | 0.3384 | 0.3381 |
| | | | Dipole moments (debyes) | | | |
| HF | 0.496 | 0.850 | 0.874 | 0.880 | 0.879 | 0.878 |
| NH₃ | 1.617 | 1.742 | 1.769 | 1.766 | 1.765 | 1.766 |
| H₂CO | 0.545 | 0.941 | 1.003 | 1.008 | 1.008 | 1.006 |

by a combination of $N$ Gaussians is denoted by STO-$N$G†. We see that although the basis STO-4G gives total energies that are about $0.25 H_\infty$ ($650\,\text{kJ mol}^{-1}$) above the STO values, energy differences such as the SCF atomization energies and one-electron properties such as dipole moments are almost exactly reproduced.

TABLE 5.8 *Minimal basis (STO-3G) and experimental ground-state geometries (Newton et al. 1970a). Distances are in ångströms ($10^{-10}$ m)*

| Molecule | Parameter | STO–3G | Exp. |
|----------|-----------|--------|------|
| $CH_4$ | $R(C–H)$ | 1.083 | 1.085 |
| $C_2H_6$ | $R(C–H)$ | 1.085 | 1.096 |
|  | $R(C–C)$ | 1.538 | 1.531 |
|  | $\theta(HCH)$ | 108.2° | 107.8° |
| $C_2H_4$ | $R(C–H)$ | 1.079 | 1.076 |
|  | $R(C–C)$ | 1.305 | 1.330 |
|  | $\theta(HCH)$ | 115.4° | 116.6° |
| $C_2H_2$ | $R(C–H)$ | 1.065 | 1.061 |
|  | $R(C–C)$ | 1.168 | 1.203 |
| $CH_3F$ | $R(C–H)$ | 1.097 | 1.105 |
|  | $R(C–F)$ | 1.384 | 1.385 |
|  | $\theta(HCH)$ | 108.3° | 109.9° |
| $CH_2F_2$ | $R(C–H)$ | 1.109 | 1.091 |
|  | $R(C–F)$ | 1.378 | 1.358 |
|  | $\theta(HCH)$ | 108.8° | 112·1° |
|  | $\theta(FCF)$ | 108.7° | 108.2° |
| $H_2O$ | $R(O–H)$ | 0.990 | 0.957 |
|  | $\theta(HOH)$ | 100.0° | 104.5° |
| $H_2CO$ | $R(C–H)$ | 1.101 | 1.101 |
|  | $R(C–O)$ | 1.217 | 1.203 |
|  | $\theta(HCH)$ | 114.5° | 116.5° |
| $H_2O_2$ | $R(O–H)$ | 1.001 | 0.950 |
|  | $R(O–O)$ | 1.396 | 1.475 |
|  | $\theta(HOO)$ | 101.1° | 94.8° |
|  | Dihedral angle | $125 \pm 2°$ | 111.5° |

Several ground-state molecular geometries obtained with the three-Gaussian basis STO-3G are compared in table 5.8 with the experimental values (Newton *et al.* 1970a; the experimental geometries are those quoted by Newton *et al.* ). The agreement with experiment is remarkably good, and in many cases is of quantitative quality. The results are fairly typical of the restricted Hartree–Fock theory, although some improvement is often obtained by an extension of the basis, particularly when there are valence lone pairs present. For example (set 9 in table 5.5), an accurate

† An *s*-type Slater function is represented by a combination of 1*s* Gaussians only, a *p*-type Slater function by 2*p* Gaussians only, and so on.

## Representation of the orbitals

SCF calculation for $H_2O$ gives $R_{OH} = 0.941$ Å and $\angle HOH = 106.6°$, whilst Dunning and Winter (1971; see also §7.3.2) have shown that a proper inclusion of polarization functions is essential for the accurate description of the internal rotation in $H_2O_2$; their calculation, using the contracted Gaussian basis of $O(9s, 4p, 1d/4s, 3p, 1d)$ and $H(4s, 1p/2s, 1p)$, gave a dihedral angle of $113.7°$ (see table 5.8). In general, however, the ground-state geometry obtained within the RHF scheme is rather insensitive to the nature and the quality of the basis.

### 5.7 DOUBLE-ZETA AND EXTENDED BASES

In contrast to the ground-state geometries, the example of the dipole moments of CO and $H_2O$ (tables 5.3 and 5.5) shows that the values of other molecular properties are often strongly basis-dependent. Properties such as the dipole moment or energies of reaction may however be expected to be insensitive to the quality of the description of atomic inner-shell orbitals, and Pople and co-workers (for example: Ditchfield, Hehre and Pople 1971; Hehre, Ditchfield and Pople 1972; Hariharan and Pople 1973) have investigated how the description of molecular properties varies with increasing size and flexibility of the basis used to represent the valence molecular orbitals, whilst keeping a minimal basis for the inner shells. Total SCF energies for several bases of this type are compared in table 5.9 with energies obtained with a double-zeta contracted Gaussian basis, $(10s, 5p/4s, 2p)$ for C to F and $H(4s/2s)$ (Snyder and Basch 1969), and with more accurate (RHF) energies. In the basis sets $N$-31G ($N = 4$ and 6) each inner-shell orbital is represented by a single basis function which is a combination of $N$ Gaussians and each atomic valence orbital (including that of hydrogen) is split into inner and outer parts described by three Gaussians and one Gaussian respectively (Ditchfield et al. 1971; Hehre et al. 1972). The basis 6–31G* is 6–31G plus a set of $d$-type Gaussian polarization functions on each atom other than hydrogen, and the basis 6–31G** has, in addition, a set of $p$-type Gaussians on each hydrogen atom. That the total energy decreases with increasing size of basis is not surprising, and the only notable result is obtained from a comparison of columns 4 and 7. The 6–31G and double-zeta (DZ) bases give very similar energies for the hydrocarbons and $H_2$, but the 6–31G basis is normally the inferior for those molecules containing valence lone pairs.

Although total molecular energies are clearly strongly dependent

TABLE 5.9 *Total SCF energies (in units of $H_\infty$)*

| Molecule | STO-3G (a) | 4-31G (b) | 6-31G (c) | 6-31G* (c) | 6-31G** (c) | DZ (d) | More accurate | |
| --- | --- | --- | --- | --- | --- | --- | --- | --- |
| | | | | | | | (e) | (f) |
| H₂ | −1.1175 | −1.1268 | −1.1268 | −1.1268 | −1.1313 | −1.1266 | −1.1336 | |
| N₂ | −107.501 | −108.747 | −108.868 | −108.942 | −108.942 | −108.870 | −108.997 | |
| O₂ | −147.634 | −149.392 | −149.545 | −149.614 | −149.614 | | | |
| F₂ | −195.982 | −198.449 | −198.646 | −198.673 | −198.673 | −198.693 | −198.770 | |
| HF | −98.573 | −99.887 | −99.983 | −100.003 | −100.011 | −100.015 | −100.071 | |
| H₂O | −74.966 | −75.908 | −75.981 | −76.010 | −76.023 | −76.004 | −76.065 | |
| NH₃ | −55.455 | −56.105 | −56.163 | −56.184 | −56.195 | −56.171 | −56.225 | |
| CH₄ | −39.727 | −40.140 | −40.180 | −40.195 | −40.202 | −40.182 | −40.225 | −40.214 |
| C₂H₂ | −75.856 | −76.711 | −76.793 | −76.817 | −76.821 | −76.792 | −76.860 | −76.848 |
| C₂H₄ | −77.074 | −77.921 | −78.003 | −78.030 | −78.038 | −78.005 | | −78.062 |
| C₂H₆ | −78.306 | −79.115 | −79.197 | −79.228 | −79.237 | −79.198 | | −79.259 |
| CH₃F | −137.169 | −138.857 | −138.992 | −139.035 | −139.040 | | | |
| CH₂O | −112.354 | −113.692 | −113.808 | −113.864 | −113.867 | −113.821 | | |
| HCN | −91.675 | −92.731 | −92.828 | −92.873 | −92.875 | −92.829 | −92.920 | |

References: (a) Lathan et al. 1974; (b) Hehre et al. 1972; (c) Hariharan and Pople 1973; (d) Snyder and Basch 1969; (e) estimated Hartree–Fock limits quoted by Hariharan and Pople 1973; (f) Clementi and Popkie 1972.

TABLE 5.10 *Hydrogenation energies* (kJ mol$^{-1}$)

| Reaction | STO-3G (a) | 4-31G (c) | 6-31G (c) | 6-31G** (c) | DZ (a) | More accurate (e) | (f) | Exp. |
|---|---|---|---|---|---|---|---|---|
| $C_2H_2 + 3H_2 \rightarrow 2CH_4$ | −644 | −493 | −493 | −493 | −506 | −498 | −468 | −441 |
| $C_2H_4 + 2H_2 \rightarrow 2CH_4$ | −381 | −276 | −273 | −271 | −278 | | −258 | −240 |
| $C_2H_6 + H_2 \rightarrow 2CH_4$ | −80 | −98 | −98 | −91 | −104 | | −92 | −75 |
| $N_2 + 3H_2 \rightarrow 2NH_3$ | −148 | −198 | −206 | −141 | −246 | −138 | | −158 |
| $HCN + 3H_2 \rightarrow CH_4 + NH_3$ | −405 | −349 | −356 | −335 | −380 | −339 | | −321 |
| $F_2 + H_2 \rightarrow 2HF$ | −122 | −497 | −510 | −573 | −551 | −623 | | −560 |
| $CH_3F + H_2 \rightarrow CH_4 + HF$ | −36 | −115 | −118 | −110 | | | | −123 |
| $O_2 + 2H_2 \rightarrow 2H_2O$ | −165 | −448 | −449 | −441 | | | | −523 |
| $CH_2O + 2H_2 \rightarrow CH_4 + H_2O$ | −273 | −269 | −273 | −248 | −294 | | | −240 |

The references (a)–(f) are as in table 5.9. The experimental energies are enthalpies of hydrogenation at o K and corrected for zero-point vibrations.

on the nature of the basis, table 5.10 shows that this is not necessarily the case for energy differences. Although we saw in §3.6 (table 3.4) that dissociation energies are generally poorly described within the RHF approximation (for example, the error in the total dissociation energy of $CH_4$ is $0.144H_\infty$ or $378\,kJ\,mol^{-1}$), the hydrogenation energies shown in table 5.10 suggest that the correlation energies of the individual reactants and products of a chemical reaction can often be expected to cancel when the molecules involved are all in closed-shell states (Snyder and Basch 1969). This general conclusion is confirmed by studies of several other types of reactions by Pople and co-workers. It is important to remember however that reliable values of energy differences can be obtained only when the same standard basis is used for all the molecules involved in a reaction, in order to allow the cancellation of errors arising from the use of a limited basis. We see (table 5.10) that this cancellation is very effective when the valence orbitals are described by a basis of about double-zeta and greater accuracy, but not for the minimal basis. A minimal description of the valence orbitals is not normally reliable, particularly when fluorine is involved, but the results obtained with the bases $N$-31G ($N = 4$ and 6) suggest that energies of reaction are insensitive to the description of the inner-shell orbitals. In all cases, the basis 4–31G gives a qualitatively correct description of the magnitudes and trends of energies of reaction, and the inclusion of polarization functions does not appear to change this description in any significant way, particularly when there are no valence lone pairs present.

## 5.8 EXCITED STATES

In a simple qualitative molecular-orbital theory, the different electronic states of a molecule are often envisaged in terms of the different ways of distributing the electrons amongst a given set of molecular orbitals. This picture is a useful first approximation for some simple states, such as those involving a single valence electron outside closed shells, but, in general, excited states are not described quite so simply. The principal factor to be taken into account, within the orbital approximation, is the relaxation of the charge distribution on going from one state to another. That is, a separate SCF calculation should be performed for each state to allow the orbitals to adapt themselves to the particular distribution of the electrons amongst the orbitals in each state. In addition, we must remember (see §3.3.1) that the virtual orbitals that are obtained from a

## Representation of the orbitals

SCF calculation for the ground state are not normally suitable for the description of excited states, in the sense that an excited configuration obtained by replacing one or more of the occupied ground-state orbitals by virtual orbitals is not normally a good approximation to an excited-state wave function. The inclusion of configuration interaction therefore becomes particularly important when an excited-state wave function is constructed from ground-state (occupied and virtual) orbitals.

The one-configuration SCF approach (as in RHF theory, for example) is suitable only when the state of interest is the lowest state of a given symmetry. The treatment of an excited state is then essentially the same as that for the ground state; for example, an SCF (RHF) calculation followed, if required, by CI or some other method of including electron correlation. Alternatively, the SCF calculation for the excited state may be simulated by using the ground-state orbitals and performing a limited CI calculation. For example, if the excited state can be considered as a 'singly excited' state, then a suitable wave function is a combination of singly substituted configurations derived from the ground-state SCF wave function. If the excited state has the same symmetry as the ground state however, or if it is not the lowest excited state of a given symmetry, then CI must be used to ensure that the wave function be orthogonal to the wave functions of all the lower states, and that the corresponding energy be an upper bound to the exact energy of the state in question.

As an example of the calculation of wave functions for several states of the same system, we consider the ground and low-lying excited states of methylene. This molecule is of interest not only because of the importance of the $CH_2$ group in organic chemistry, but also because its ground state was until 1970 generally accepted to be linear (Herzberg 1967). A good CI calculation by Foster and Boys in 1960, however, gave a bond angle of $129°$ for the $^3B_1$ ground state, and the non-linear geometry has been confirmed by recent more accurate theoretical work (Bender and Schaefer 1970; McLaughlin, Bender and Schaefer 1972; Staemmler 1973), and by new experimental studies (Bernheim et al. 1970; Wasserman, Yager and Kuck 1970; Herzberg and Johns 1971). The angle determined experimentally by Wasserman et al. (1970) is $136 \pm 8°$.

The results of a set of SCF and CI calculations for the ground state of $CH_2$ are compared in table 5.11 with those of comparable calculations for the ground state of $H_2O$ (McLaughlin et al. 1972). The basis used in these is an extended contracted Gaussian basis, ($10s$, $6p$, $2d/5s$, $3p$, $1d$) for C and O, and ($5s$, $3p/3s$, $1p$) for H, which gives SCF results for $H_2O$

TABLE 5.11 *Total energies, geometries and force constants for the ground states of* $H_2O$ *and* $CH_2$ *(McLaughlin, Bender and Schaefer 1972)*

| Property | $H_2O$ | | | $CH_2$ | | |
|---|---|---|---|---|---|---|
| | SCF | CI | Exp. | SCF | CI | Exp. |
| $-E/H_\infty$ | 76.0588 | 76.1738 | | 38.9327 | 39.0121 | |
| $R/a_\infty$ | 1.780 | 1.829 | 1.809 | 2.026 | 2.056 | 2.037† |
| $\angle$HXH | 105.8° | 103.4° | 104.5° | 129.5° | 134.0° | $136\pm8°$‡ |
| $F_r/N$ cm$^{-1}$ | 8.72 | 8.75 | 8.4 | 6.16 | 6.13 | |
| $F_\theta/N$ cm$^{-1}$ | 0.88 | 0.83 | 0.76 | 0.44 | 0.33 | |

The bond stretching and bond bending force constants, $F_r$ and $F_\theta$, have been obtained by fitting the energy to $2E = F_r(r_1^2+r_2^2)+F_\theta R^2\theta^2$, where $r_1$ and $r_2$ are the bond stretches, $\theta$ is the bond angle distortion, and $R$ is the equilibrium bond length.
    † Herzberg and Johns 1971.   ‡ Wasserman *et al.* 1970.

close to the RHF limit (see table 5.5). The CI expansions contain several hundred singly and doubly substituted configurations obtained by replacing one or two valence orbitals in the SCF configuration by excited orbitals. The results of the $H_2O$ calculations suggest that the CI bond length ($2.056a_\infty$) of $CH_2$ is probably accurate to within $0.03a_\infty$ and the bond angle ($134°$) to within $2°$. An independent calculation by Staemmler (1973), using a correlated pair function approach (§4.3), has given very similar results for the geometry and force constants of $CH_2$. The experimentally determined values shown in table 5.11 are less accurate.

O'Neill, Schaefer and Bender (1971) have used a double-zeta basis of contracted Gaussians, C($9s, 5p/4s, 2p$) and H($4s/2s$), for SCF and CI calculations of the potential energy surfaces of the lowest $^1A_1$, $^1A_2$, $^3A_2$, $^1B_1$, $^1B_2$, and $^3B_2$ excited states of $CH_2$, as well as of the $^3B_1$ ground state. The CI wave functions for these states are dominated by the following orbital configurations, which are also the SCF configurations:

$$(1a_1)^2(2a_1)^2(1b_2)^2(3a_1)(1b_1) \quad {}^1B_1, {}^3B_1$$
$$(1a_1)^2(2a_1)^2(1b_2)^2(3a_1)^2 \quad {}^1A_1$$
$$(1a_1)^2(2a_1)^2(1b_2)(3a_1)^2(1b_1) \quad {}^1A_2, {}^3A_2$$
$$(1a_1)^2(2a_1)^2(1b_2)(3a_1)(1b_1)^2 \quad {}^1B_2, {}^3B_2$$

Only the three lowest states, $^3B_1$, $^1A_1$ and $^1B_1$, are predicted to be bound, the others being unstable with respect to dissociation to a carbon atom and a hydrogen molecule. The results for the three bound states are shown in table 5.12, and they show that the geometries of low-lying excited states are as well reproduced by the single-configuration SCF

TABLE 5.12   *Total energies, geometries, and force constants for the lowest* $^3B_1$, $^1A_1$ *and* $^1B_1$ *states of* $CH_2$
*(O'Neil, Schaefer and Bender 1971)*

| Property | $^3B_1$ | | $^1A_1$ | | | $^1B_1$ | | |
|---|---|---|---|---|---|---|---|---|
| | SCF | CI | SCF | CI | Exp.† | SCF | CI | Exp.† |
| $-E/H_\infty$ | 38.9136 | 38.9826 | 38.8620 | 38.9472 | | 38.8452 | 38.9114 | |
| $R_{OH}/a_\infty$ | 2.031 | 2.069 | 2.085 | 2.142 | 2.10 | 2.017 | 2.063 | 1.98 |
| ∠ HCH | 130.4° | 133.3° | 106.5° | 104.4° | 102.4° | 150.5° | 143.8° | 140 |
| $F_r/N$ cm$^{-1}$ | 5.06 | 5.06 | 5.18 | 5.09 | | 6.94 | 5.00 | |
| $F_\theta/N$ cm$^{-1}$ | 0.37 | 0.29 | 0.62 | 0.53 | | 0.36 | 0.15 | |

† Herzberg and Johns 1966.

theory as are ground-state geometries. The SCF bond lengths obtained with the double-zeta basis are closer to the experimental values than are the CI values, but the inclusion of polarization functions characteristically leads to shorter SCF bond lengths, approaching RHF values which are nearly always shorter than the observed values. On the other hand, O'Neil *et al.* (1971) conclude, from this and other calculations, that the double-zeta CI bond lengths should be between 0.04 and $0.06a_\infty$ longer than the experimental values. The experimental bond length for the $^1B_1$ state is therefore probably about $0.03a_\infty$ longer than the value ($1.98a_\infty$) quoted in table 5.12.

A property of methylene which is of some interest because of the importance of carbenes as intermediates in many organic reactions is the magnitude of the singlet–triplet ($^1A_1$–$^3B_1$) energy separation (see for example, Bethell 1973). This quantity cannot be determined directly from experiment, but indirect estimates vary from 4–10 kJ mol$^{-1}$ from analysis of the photolysis of ketene (Halberstadt and McNesby 1967; Carr, Eder and Topor 1970) to an upper bound of about 100 kJ mol$^{-1}$ from spectroscopic studies (Herzberg 1967). The results of a number of non-empirical calculations are summarized in table 5.13. Estimates of

TABLE 5.13 *Singlet–triplet ($^1A_1$–$^3B_1$) energy separation in* $CH_2$

| Basis | $-E/H_\infty$ | | $\Delta E$/kJ mol$^{-1}$ |
|---|---|---|---|
| | $^3B_1$ | $^1A_1$ | |
| $C(9s, 5p/4s, 2p)$, $H(4s/2s)$ (O'Neil *et al.* 1971) | 38.9136 (1-config. SCF) | 38.8620 (1-config. SCF) | 135.5 (1) |
| | 38.9136 (1-config. SCF) | 38.8772 (2-config. SCF) | 95.6 (2) |
| | 38.9826 (CI) | 38.9472 (CI) | 92.9 (3) |
| $C(10s, 5p, 2d/5s, 3p, 1d)$, $H(5s, 3p/3s, 1p)$ (McLaughlin *et al.* 1972; Bender *et al.* 1972) | 38.9327 (1-config. SCF) | 38.9140 (2-config. SCF) | 49.1 (4) |
| | 39.0121 (CI) | 38.9898 (CI) | 58.5 (5) |
| $C(11s, 7p, 2d/6s, 4p, 2d)$, $H(5s, 2p/3s, 2p)$ (Staemmler 1973) | 38.9308 (1-config. SCF) | 38.8909 (1-config. SCF) | 104.8 (6) |
| | 39.0754 (Correlated pairs) | 39.0607 (Correlated pairs) | 38.6 (7) |

# Representation of the orbitals

the RHF and exact splitting energies (Bender *et al.* 1972; Staemmler 1973) suggest that the RHF value lies in the range $80$–$100\,\mathrm{kJ\,mol^{-1}}$, and that the exact value lies in the range $25$–$55\,\mathrm{kJ\,mol^{-1}}$, with a probable value near $40\,\mathrm{kJ\,mol^{-1}}$. The large RHF value (see also (1) and (6) in table 5.13) shows that electron correlation is more important in the excited ($^1A_1$) state than it is in the ground state, and this may be understood in terms of the creation of a new electron pair in the excited state: $(3a_1)(1b_1) \rightarrow (3a_1)^2$. The values (2) and (4) in table 5.13 show, however, that most of the additional correlation energy of the excited state can be recovered by performing a two-configuration SCF calculation. The two configurations involved are

$$(1a_1)^2(2a_1)^2(1b_2)^2(3a_1)^2 \quad \text{and} \quad (1a_1)^2(2a_1)^2(1b_2)^2(1b_1)^2$$

and the resulting wave function (ignoring all but the lone-pair orbitals) has the form

$$|3a_1\alpha, 3a_1\beta| - c^2|1b_1\alpha, 1b_1\beta|, \quad c^2 \approx 0.2 \qquad (5.16)$$

The effect of adding the second configuration is to separate the two electrons, and is equivalent to a relaxation of the double-occupancy and orbital-symmetry constraints of RHF theory similar to that discussed in §4.1 for the angular correlation in the ground state of helium. Thus (apart from normalization) (5.16) can be written in the extended Hartree–Fock form

$$\mathscr{P}_{^1A_1}|\psi\alpha, \psi'\beta|$$

where $\mathscr{P}_{^1A_1}$ is the $^1A_1$ symmetry projection operator, and

$$\psi = 3a_1 + c1b_1, \quad \psi' = 3a_1 - c1b_1$$

are new lone-pair orbitals directed above and below the molecular plane and away from the hydrogens.

We can conclude from this discussion that the (one-configuration) RHF model does not in general provide a satisfactory description of the energy separations between different states of a molecule. Furthermore, a comparison of the singlet–triplet splitting for the different basis sets in table 5.13 shows that this quantity is strongly dependent on the basis. The inclusion of $d$-type polarization functions on the carbon is particularly important in both the single- and multi-configuration wave functions (Bender *et al.* 1972; Staemmler 1973). The inclusion of these functions affects mainly the $3a_1$ orbital, and is therefore about twice as effective in lowering the energy of the $^1A_1$ state as of the $^3B_1$ state. A limited basis cannot therefore be expected to provide reliable results even at the RHF level of approximation.

**5.9** THE FLOATING SPHERICAL GAUSSIAN ORBITAL MODEL

The basis sets considered so far have all been made up of nuclear-centred functions; that is, functions of the type (5.10) whose centres are fixed on the nuclei of a molecule. Molecular orbitals constructed from Gaussian functions may, however, be given additional variational freedom if the centres of the Gaussians are themselves treated as variational parameters, and the functions are allowed to 'float' during the variational optimization of the energy. For example, a molecular orbital which is expressed as a linear combination of $n$ $1s$-type Gaussians then contains $5n - 1$ variational parameters: the $n - 1$ independent coefficients, the $n$ Gaussian exponents, and the $3n$ position coordinates of the centres. All reference to the atomic orbitals of the constituent atoms of a molecule is lost in such a description, and the use, for example, of polarization functions to describe the distortion of the atomic orbitals in the molecular environment is no longer necessary or relevant.

Although a basis of floating Gaussians can clearly be used to obtain highly accurate molecular wave functions, Frost (1967) has shown how such a basis can also be used to set up a very simple orbital model of molecular structure in which the molecular orbitals are expressed in terms of the smallest possible basis set consistent with the number of electrons in a molecule. In this model, the floating spherical Gaussian orbital (FSGO) model, the electronic structure of a molecule in a closed-shell state is described by $N/2$ doubly occupied localized bond and lone-pair orbitals, each of which is represented by a *single* normalized spherical ($1s$) Gaussian

$$\chi = (2/\pi\rho^2)^{\frac{3}{4}} \exp\left(-|r - R|^2/\rho^2\right)$$

whose effective radius $\rho$ (or exponent $1/\rho^2$) and position $R$ are determined by minimization of the energy. For example, only two spherical Gaussian orbitals (SGO) are required to accommodate the four electrons in the ground state of LiH, and these may be assumed to have their centres on the molecular axis. In the methane molecule, the ten electrons are described by five SGO, with the carbon inner shell represented by one SGO and each of the four equivalent C–H bond pairs by a single SGO centred on a bond axis. Because of the simplicity of the basis, the simple FSGO model cannot be expected to provide very accurate total or orbital energies. Thus total energies are typically only about 85 per cent of the accurate RHF values, and the discrepancy is due mainly to the unsatisfactory description of the inner shell orbitals. On the other hand,

TABLE 5.14 *FSGO and experimental ground-state geometries. Distances are in ångströms* $(10^{-10} \text{ m})$

| Molecule | Reference | Parameter | FSGO | Exp. |
|---|---|---|---|---|
| $CH_4$ | a | $R(C–H)$ | 1.115 | 1.085 |
| $C_2H_6$ | b | $R(C–H)$ | 1.120 | 1.096 |
| | | $R(C–C)$ | 1.501 | 1.531 |
| | | $\theta(HCH)$ | 108.2° | 107.8° |
| $C_2H_4$ | b | $R(C–H)$ | 1.101 | 1.076 |
| | | $R(C–C)$ | 1.351 | 1.330 |
| | | $\theta(HCH)$ | 118.7° | 116.6° |
| $C_2H_2$ | b | $R(C–H)$ | 1.079 | 1.061 |
| | | $R(C–C)$ | 1.214 | 1.203 |
| LiH | c | $R(Li–H)$ | 1.707 | 1.595 |
| $NH_3$ | a | $R(N–H)$ | 1.011 | 1.012 |
| | | $\theta(HNH)$ | 87.6° | 106.6° |
| $H_2O$ | a | $R(O–H)$ | 0.881 | 0.957 |
| | | $\theta(HOH)$ | 88.4° | 104.5° |
| $B_2H_6$ | d | $R(B–B)$ | 1.84 | 1.77 |
| | | $R(B–H_t)$ | 1.24 | 1.19 |
| | | $R(B–H_b)$ | 1.38 | 1.34 |
| | | $\theta(BBH_t)$ | 117.2° | 119.9° |
| | | $\theta(H_bBH_b)$ | 96.7° | 97.0° |
| | | $\theta(H_tBH_t)$ | 125.6° | 120.2° |

References: (a) Frost 1968; (b) Frost and Rouse 1968; (c) Frost 1967; (d) Frost 1970.

molecular geometries, as shown in table 5.14, are surprisingly well described, particularly for non-polar molecules. The large errors in the bond angles of $NH_3$ and $H_2O$ are probably due to an inadequate description of the lone pairs in these molecules. In $H_2O$ the two SGO representing the lone pairs have a strong tendency to coalesce during the variational minimization of the energy, and must therefore be constrained to stay apart (Frost 1968). A similar problem occurs with the double bond in $C_2H_4$ which is described by two identical SGO at equal distances above and below the molecular plane at the mid-point of the C–C axis (Frost and Rouse 1968).

The simple FSGO model involves the optimization of up to four parameters for each independent SGO, and is a practical procedure only for small molecules. Christoffersen, Genson and Maggiora (1971; see also Christoffersen 1972) have proposed a modification of the Frost model which is more suitable for large molecules. Such a molecule is regarded as built up of smaller fragments, for which FSGO calculations are performed, and the resulting SGO are then combined to form a fixed basis for a subsequent conventional molecular-orbital treatment of the whole molecule. The fragments need not correspond to stable molecular species, and many molecules can be built up from quite a small number of fragments. Thus, an FSGO calculation for $CH_4$ provides a set of SGO suitable for the description of saturated hydrocarbons. Other fragments include planar $CH_3$ for unsaturated hydrocarbons, tetrahedral $NH_3$ for amines, planar $NH_3$ for amides and pyrrole, 'tetrahedral' $H_2O$ (with a tetrahedral arrangement of the valence SGO) for esters and alcohols, and 'planar' $H_2O$ (with one lone pair occupying a $\pi$ orbital on the oxygen) for furan. In planar $CH_3$, $NH_3$, and $H_2O$ the atomic $\pi$ orbital is represented as a combination of two identical SGO (lobes) at equal distances above and below the plane. The essential differences between the Christoffersen and Frost models are (a) that the basis in the Christoffersen model is not fully optimized for the whole molecule, and (b) that each inter-fragment bond is described by two SGO, one from each fragment, instead of the single SGO of the Frost model. The two models however give very similar energies and geometries.

Christoffersen and co-workers have demonstrated the viability of the FSGO fragment approach by numerous calculations on small and medium-sized organic molecules, including studies of the relative stabilities of a number of benzene and naphthalene isomers (Christoffersen 1971), of anthracene and phenanthrene (Christoffersen 1973), and of several conformations of the mono- to penta-peptides of glycine (Shipman and Christoffersen 1973). Erickson and Linnett (1972) have also shown that the FSGO approach can be applied to very large systems, by a calculation of the energy and structure of crystalline LiH. They obtain an internuclear distance of 1.978 Å which is only 3 per cent smaller than the experimental value of 2.043 Å (the value 1.710 Å for the isolated molecule is 7 per cent larger than the experimental distance), and an energy of sublimation of 273 kJ mol$^{-1}$ which is 24 per cent larger than the experimental value (corrected to 0 K) of 220 kJ mol$^{-1}$.

The Frost and Christoffersen FSGO models, as well as Pople's STO–*N*G model, provide simple non-empirical methods for the description of the shapes and electronic structures of molecules which, in accuracy and ease of computation, rival the several sophisticated semi-empirical methods currently widely used for large molecules (Hoffmann 1963; Dewar 1969; Pople and Beveridge 1970). They have two clear advantages over semi-empirical theories: (a) they are non-empirical theories which can

## Representation of the orbitals

be applied to an arbitrary configuration of any molecule, for which experimental information may or may not be available, without the empirical parametrization and calibration which the semi-empirical theories require, and (*b*) they are in principle capable of improvement to give any desired accuracy, within and beyond the orbital approximation.

# 6 The electron distribution

From the electronic wave function of a stationary state of an $N$-electron system it is possible to calculate those properties of the system which depend primarily on the electronic structure in this state. Such calculations provide quantitative tests of the relative merits of different models and approximations, and they serve to supplement and check the results of experiment. The computation of molecular wave functions also provides a basis for the interpretation of the results of experimental observation, and it is to the interpretative role of computational quantum chemistry that we now turn our attention. The present chapter is devoted to the analysis and interpretation of electronic wave functions.

An $N$-electron wave function involves the $3N$ space coordinates and the $N$ spin coordinates of the electrons, and it may in general be a correlated wave function expressed as a linear combination of many Slater determinants, constructed from orbitals which may in turn be linear combinations of basis functions. The interpretation of such a wave function can clearly be a formidable problem, but, fortunately, the essential physical content of a wave function is embodied in a small number of rather simpler derived quantities, the one-electron and two-electron density matrices and density functions (excellent accounts of these are given by Löwdin 1959$a$ and by McWeeny and Sutcliffe 1969, chapter 4). The most important of these derived quantities for interpretative purposes is the one-electron density function. This is a function of one set of four space-spin coordinates only, and it provides information about the overall spatial distribution of electronic charge and spin.

## 6.1 THE ONE-ELECTRON DENSITY FUNCTION

We consider first a single electron in the normalized orbital $\psi(\boldsymbol{r})$. The physical interpretation of the orbital is that

$$\psi(\boldsymbol{r})\,\psi(\boldsymbol{r})^*\,\mathrm{d}v = |\psi(\boldsymbol{r})|^2\,\mathrm{d}v$$

is the probability of finding the electron in the volume element $\mathrm{d}v$ at position $\boldsymbol{r}$. The quantity $|\psi(\boldsymbol{r})|^2$ is a charge-density function which describes the spatial distribution of electronic charge in the system: $-e|\psi(\boldsymbol{r})|^2\,\mathrm{d}v$ is the amount of electronic charge to be found in the

## The electron distribution

volume element $dv$ at position $\boldsymbol{r}$. If the electron has spin $m_s = +\tfrac{1}{2}$ then the corresponding wave function is the spin-orbital $\phi(\boldsymbol{r}, \sigma) = \psi(\boldsymbol{r})\alpha(\sigma)$, and the quantity

$$|\phi(\boldsymbol{r}, \sigma)|^2 d\tau = |\psi(\boldsymbol{r})|^2 dv \cdot |\alpha(\sigma)|^2 d\sigma$$

is interpreted as the probability of finding the electron in $dv$ with a value of the spin variable in the range $\sigma$ to $\sigma + d\sigma$. The interpretation is made complete if we assume that the spin function $\alpha(\sigma)$ vanishes at all values of $\sigma$ except $\sigma \approx +\tfrac{1}{2}$ (McWeeny 1972, p. 102). Such a function may be envisaged as a sharp 'spike' of vanishing width in the vicinity of $\sigma = +\tfrac{1}{2}$ as illustrated in fig. 6.1, while $\beta(\sigma)$ is a similar spike in the vicinity of $\sigma = -\tfrac{1}{2}$. The normalization condition

$$\int |\alpha(\sigma)|^2 d\sigma = 1$$

means that each spike has unit area.

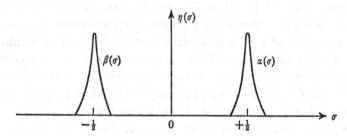

Fig. 6.1. Schematic representation of spin functions. $\alpha(\sigma)$ and $\beta(\sigma)$ are regarded formally as functions of the spin variable $\sigma$, having sharp 'spikes' of vanishing width at $\sigma = +\tfrac{1}{2}$ and $\sigma = -\tfrac{1}{2}$ respectively (after McWeeny 1972, p. 102).

For a two-electron system in the state with normalized wave function $\Psi(\boldsymbol{r}_1, \sigma_1, \boldsymbol{r}_2, \sigma_2)$, the quantity

$$|\Psi(\boldsymbol{r}_1, \sigma_1, \boldsymbol{r}_2, \sigma_2)|^2 d\tau_1 d\tau_2$$

is interpreted as the simultaneous probability of finding one electron (labelled 1) with coordinates in the space-spin volume element $d\tau_1 = dv_1 d\sigma_1$ and the second electron with coordinates in the element $d\tau_2 = dv_2 d\sigma_2$. The probability of finding electron 1 in $d\tau_1$ irrespective of the position and spin of electron 2 is then obtained by integrating over the coordinates of electron 2:

$$\left[\int |\Psi(\mathbf{r}_1, \sigma_1, \mathbf{r}_2, \sigma_2)|^2 \, d\tau_2\right] d\tau_1$$

The probability of finding *either* electron in $d\tau_1$ is twice this since the two-electron density $|\Psi|^2$ is symmetric in the coordinates of the two electrons. Thus

$$\left[2\int |\Psi(\mathbf{r}_1, \sigma_1, \mathbf{r}_2, \sigma_2)|^2 \, d\tau_2\right] d\tau_1 = \rho(\mathbf{r}_1, \sigma_1) \, d\tau_1$$

= probability of finding either electron in the volume element $d\tau_1 = dv_1 \, d\sigma_1$ at $(\mathbf{r}_1, \sigma_1)$, independent of the position and spin of the other electron.

The quantity
$$\rho(\mathbf{r}_1, \sigma_1) = 2\int |\Psi(\mathbf{r}_1, \sigma_1, \mathbf{r}_2, \sigma_2)|^2 \, d\tau_2$$

is the *one-electron density function* of the two-electron system in the state with wave function $\Psi$.

The generalization to the case of an $N$-electron wave function is straightforward. Thus

$$|\Psi(\mathbf{r}_1, \sigma_1, \mathbf{r}_2, \sigma_2, \ldots, \mathbf{r}_N, \sigma_N)|^2 \, d\tau_1 \, d\tau_2 \ldots d\tau_N$$

is the simultaneous probability of finding one electron (labelled 1) with coordinates in the space–spin volume element $d\tau_1$, electron 2 in $d\tau_2$, and so on. The corresponding one-electron density function is

$$\rho(\mathbf{r}_1, \sigma_1) = N\int |\Psi(\mathbf{r}_1, \sigma_1, \mathbf{r}_2, \sigma_2, \ldots, \mathbf{r}_N, \sigma_N)|^2 \, d\tau_2 \ldots d\tau_N \qquad (6.1)$$

and $\rho(\mathbf{r}, \sigma) \, d\tau$ is the probability of finding *any one* of the $N$ electrons in the element $d\tau = dv \, d\sigma$ at $(\mathbf{r}, \sigma)$, independent of the positions and spins of the other electrons.

The one-electron density function has a particularly simple form in the orbital approximation when the wave function is a single Slater determinant of orthonormal spin-orbitals. Thus, for a two-electron system,

$$\Psi(1, 2) = |\phi_1(1), \phi_2(2)| = (1/\sqrt{2})[\phi_1(1)\,\phi_2(2) - \phi_1(2)\,\phi_2(1)]$$

and
$$\rho(\mathbf{r}_1, \sigma_1) = 2\int |\Psi(1, 2)|^2 \, d\tau_2 = |\phi_1(\mathbf{r}_1, \sigma_1)|^2 + |\phi_2(\mathbf{r}_1, \sigma_1)|^2 \qquad (6.2)$$

The total one-electron density function is therefore the sum of the

# The electron distribution

individual spin-orbital density functions. If, for example, the spin-orbitals have different spin factors, $\phi_1 = \psi^\alpha \alpha$ and $\phi_2 = \psi^\beta \beta$, then (6.2) can be written in the general form

$$\rho(\mathbf{r}, \sigma) = P^\alpha(\mathbf{r})|\alpha(\sigma)|^2 + P^\beta(\mathbf{r})|\beta(\sigma)|^2 \qquad (6.3)$$

in which $P^\alpha(\mathbf{r}) = |\psi^\alpha(\mathbf{r})|^2$ and $P^\beta(\mathbf{r}) = |\psi^\beta(\mathbf{r})|^2$ are spinless densities associated with the spins $+\frac{1}{2}$ and $-\frac{1}{2}$ respectively. The quantity $P^\alpha(\mathbf{r})\,\mathrm{d}v$ is interpreted as the probability of finding an electron with spin $m_s = +\frac{1}{2}$ in the volume element $\mathrm{d}v$ at position $\mathbf{r}$.

The density function $\rho(\mathbf{r}, \sigma)$ of *any* $N$-electron wave function, which has a well-defined value of the total spin $M_S$, can always be expressed in the form (6.3), as the sum of separate contributions from the two spin components. $P^\alpha$ and $P^\beta$ describe the electron charge distributions associated with the two types of spin, and the total charge density, which can be obtained from $\rho$ by integration over the spin, is the sum of these:†

$$P(\mathbf{r}) = \int \rho(\mathbf{r}, \sigma)\,\mathrm{d}\sigma = P^\alpha(\mathbf{r}) + P^\beta(\mathbf{r}) \qquad (6.4)$$

This is the probability per unit volume of finding any electron of either spin in $\mathrm{d}v$ at $\mathbf{r}$, independent of the positions and spins of the other electrons, and it is the quantity which is observed, for example, in X-ray crystallography. A second quantity which can be derived from $\rho$ is the spin density

$$P^\alpha(\mathbf{r}) - P^\beta(\mathbf{r})$$

which is a measure of the excess of density of spin $+\frac{1}{2}$ over that of spin $-\frac{1}{2}$. The spin density vanishes for a singlet state, with $S = 0$, but for a non-singlet state its value at a nucleus determines the electron–nucleus contact coupling term which is observed in electron spin resonance spectroscopy (McWeeny and Sutcliffe 1969, chapter 8).

## 6.1.1 *Natural orbitals*

The one-electron density function of an arbitrary $N$-electron wave function which is a linear combination of Slater determinants constructed from a common basis set of spin orbitals $\phi_n$ has the general form

$$\rho(\mathbf{r}, \sigma) = \sum_{m,\,n} \phi_m(\mathbf{r}, \sigma) P_{mn} \phi_n(\mathbf{r}, \sigma)^* \qquad (6.5)$$

The numbers $P_{mn}$ are determined by the coefficients of the determinants

† Alternative symbols for the charge density are $\rho(\mathbf{r})$ and $P_1(\mathbf{r})$.

in the expansion of the wave function, and they are the elements of a hermitian matrix, with the property $P_{nm}^* = P_{mn}$. This matrix can be diagonalized by a suitable linear transformation of the spin-orbitals; that is, it is possible to find a new set of orthonormal spin-orbitals $\lambda_k(\mathbf{r}, \sigma)$,

$$\lambda_k(\mathbf{r}, \sigma) = \sum_n C_{nk} \phi_n(\mathbf{r}, \sigma)$$

such that the density function (6.5) is reduced to the diagonal form

$$\rho(\mathbf{r}, \sigma) = \sum_k n_k \lambda_k(\mathbf{r}, \sigma) \lambda_k(\mathbf{r}, \sigma)^* \tag{6.6}$$

in which the original functions $\phi_n$ have been replaced by the new functions $\lambda_k$, and the matrix with elements $P_{mn}$ has been replaced by the diagonal matrix with elements $n_k$, where

$$P_{mn} = \sum_k n_k C_{mk} C_{nk}^*$$

The new spin-orbitals are called the *natural spin-orbitals* of the wave function (Löwdin 1959a; Davidson 1972). The numbers $n_k$ are interpreted as occupation numbers of the spin-orbitals, and they have the property

$$0 \leqslant n_k \leqslant 1, \quad \sum_k n_k = N$$

which is an expression of the Pauli principle, that a spin-orbital may be occupied by not more than one electron.

The one-electron density function of a single determinant of $N$ orthonormal spin-orbitals is already in diagonal form

$$\rho = \sum_{n=1}^{N} \phi_n \phi_n^* \tag{6.7}$$

with each occupied spin-orbital having unit occupation number. The diagonal form of $\rho$ is particularly useful for the analysis and interpretation of the electron distribution, since $\rho$ and those properties of the system that are determined by $\rho$ are then expressed as sums of separate orbital contributions.

### 6.1.2    *The density function in Hartree–Fock theory*

Consider an $N$-electron wave function which is a linear combination of normalized Slater determinants, each of which is constructed from a common basis of orthonormal spin-orbitals $\phi_n$:

$$\Psi = C_0 \Psi_0 + \sum_{I>0} C_I \Psi_I \tag{6.8}$$

## The electron distribution

where
$$\Psi_0 = |\phi_1, \phi_2, ..., \phi_N|$$

is a suitable orbital approximation for the state in question. The corresponding one-electron density function is

$$\rho(r, \sigma) = |C_0|^2 \rho_{00}(r, \sigma) + \sum_{I>0} [C_0 C_I^* \rho_{0I}(r, \sigma) + C_0^* C_I \rho_{I0}(r, \sigma)]$$
$$+ \sum_{I,J>0} C_I C_J^* \rho_{IJ}(r, \sigma) \quad (6.9)$$

where, for example,

$$\rho_{0I}(r_1, \sigma_1) = N \int \Psi_0(r_1, \sigma_1, ..., r_N, \sigma_N) \Psi_I^*(r_1, \sigma_1, ..., r_N, \sigma_N) \, d\tau_2 ... d\tau_N$$

It can be shown that a 'transition density' like $\rho_{0I}$ is zero if the determinants $\Psi_0$ and $\Psi_I$ differ in more than one spin-orbital. As an example, for a two-electron system, let

$$\Psi_0 = |\phi_1, \phi_2|, \quad \Psi_I = |\phi_3, \phi_4|$$

Then

$$\rho_{0I}(1) = 2 \int \Psi_0(1, 2) \Psi_I^*(1, 2) \, d\tau_2$$

$$= \int [\phi_1(1) \phi_2(2) - \phi_2(1) \phi_1(2)] [\phi_3^*(1) \phi_4^*(2) - \phi_4^*(1) \phi_3^*(2)] \, d\tau_2$$

$$= \phi_1(1) \phi_3^*(1) \langle \phi_4 | \phi_2 \rangle + \phi_2(1) \phi_4^*(1) \langle \phi_3 | \phi_1 \rangle$$
$$- \phi_1(1) \phi_4^*(1) \langle \phi_3 | \phi_2 \rangle - \phi_2(1) \phi_3^*(1) \langle \phi_4 | \phi_1 \rangle \quad (6.10)$$

where
$$\langle \phi_m | \phi_n \rangle = \int \phi_m^* \phi_n \, d\tau = \delta_{mn}$$

We can consider three distinct cases:

(i) $\Psi_0$ and $\Psi_I$ are identical, with $\phi_1 = \phi_3$ and $\phi_2 = \phi_4$. Then (6.10) reduces to
$$\rho_{0I} = \rho_{00} = \phi_1 \phi_1^* + \phi_2 \phi_2^*$$
in agreement with (6.7).

(ii) $\Psi_0$ and $\Psi_I$ differ in one spin-orbital with $\phi_1 = \phi_3$ but $\phi_2 \neq \phi_4$. Then
$$\rho_{0I} = \phi_2 \phi_4^*$$

(iii) $\Psi_0$ and $\Psi_I$ differ in both spin-orbitals. Then $\rho_{0I} = 0$, since all the integrals $\langle \phi_m | \phi_n \rangle$ in (6.10) are zero.

This property of the density function, that $\rho_{0I}$ is zero if $\Psi_0$ and $\Psi_I$ differ in more than one spin-orbital, is of particular significance when

116

$\Psi_0$ is a Hartree–Fock wave function. We saw in §4.3 that the wave function to first order in a perturbative treatment of electron correlation does not contain any singly substituted configurations when the zeroth-order wave function is a Hartree–Fock wave function for which Brillouin's theorem (§3.5) is valid. The wave function to first order therefore has the form (6.8), with $\Psi_0$ as the zeroth-order term and $\sum_I C_I \Psi_I$, containing doubly substituted configurations only, as the first-order correction. The corresponding density function (6.9) then contains no terms $\rho_{0I}$, and

$$\rho = |C_0|^2 \rho_{00} + \sum_{I,J>0} \sum C_I C_J^* \rho_{IJ}$$

$$= \rho_{00} + \text{second-order and higher-order terms}$$

($|C_0|^2 = 1$ to first order). The first-order correction to the density function is therefore zero and, to first order, the Hartree–Fock density is identical to the exact density for the state in question. It is for this reason that the Hartree–Fock model is expected to provide an accurate description of the electron distribution, particularly for closed-shell states, as well as of those properties of the system, the one-electron properties, which are determined wholly by $\rho$. The example of the dipole moment (§3.5.1) suggests however, that this latter expectation is not always justified.

### 6.2  ONE-ELECTRON PROPERTIES

The value of an observable property of a quantal system is defined as the expectation value of the quantum-mechanical operator representing the property, with respect to the wave function of the state in question. If $\mathscr{F}$ is the operator representing the property $F$, its expectation value in the state with normalized wave function $\Psi$ is

$$\langle F \rangle = \int \Psi^* \mathscr{F} \Psi \, d\tau$$

and this is the observed value of the property $F$ in the state $\Psi$. If the classical expression of the property $F$ of an $N$-electron system is a function of the positions and momenta of the electrons,

$$F = F(r_1, r_2, ..., r_N; p_1, p_2, ..., p_N)$$

then the corresponding operator $\mathscr{F}$ is obtained from $F$ by replacing each momentum variable $p_j$ in $F$ by the differential operator $(\hbar/i)(\partial/\partial q_j)$ where $q_j$ is the coordinate conjugate to $p_j$ (see §2.6.2 for angular momentum).

## The electron distribution

As an example, consider a stationary state of a molecule (in the Born–Oppenheimer approximation) for which the constant energy $E$ in classical mechanics is given, as a function of the positions and momenta of the electrons, by the Hamilton function $H$:

$$E = H(\boldsymbol{r}_1, \boldsymbol{r}_2, ..., \boldsymbol{r}_N; \boldsymbol{p}_1, \boldsymbol{p}_2, ..., \boldsymbol{p}_N)$$
$$= T(\boldsymbol{p}_1, \boldsymbol{p}_2, ..., \boldsymbol{p}_N) + V(\boldsymbol{r}_1, \boldsymbol{r}_2, ..., \boldsymbol{r}_N)$$

where $T$ and $V$ are the kinetic and potential energies respectively. The kinetic energy is a function of the momenta of the electrons only,

$$T = \frac{1}{2m_e} \sum_{i=1}^{N} (p_{x_i}^2 + p_{y_i}^2 + p_{z_i}^2)$$

and the corresponding kinetic-energy operator is

$$\mathcal{T} = \frac{-\hbar^2}{2m_e} \sum_{i=1}^{N} \left( \frac{\partial^2}{\partial x_i^2} + \frac{\partial^2}{\partial y_i^2} + \frac{\partial^2}{\partial z_i^2} \right) = \frac{-\hbar^2}{2m_e} \sum_{i=1}^{N} \nabla_i^2$$

The potential energy $V$, on the other hand, is a function of the position coordinates only,

$$V = -\sum_{i=1}^{N} \sum_{\alpha=1}^{\nu} \frac{Z_\alpha e^2}{4\pi\epsilon_0 r_{i\alpha}} + \sum_{i>j=1}^{N} \frac{e^2}{4\pi\epsilon_0 r_{ij}} + \sum_{\alpha>\beta=1}^{\nu} \frac{Z_\alpha Z_\beta e^2}{4\pi\epsilon_0 R_{\alpha\beta}} \quad (6.11)$$

and is therefore its own operator. The total energy operator is then the Hamiltonian $\mathcal{H} = \mathcal{T} + V$, and the energy of the system in the stationary state with normalized wave function $\Psi$ is the expectation value of $\mathcal{H}$:

$$E = \langle \mathcal{H} \rangle = \int \Psi^* \mathcal{H} \Psi \, d\tau$$

Another example is provided by the electric dipole moment, whose classical expression

$$\boldsymbol{\mu} = e \sum_{\alpha=1}^{\nu} Z_\alpha \boldsymbol{R}_\alpha - e \sum_{i=1}^{N} \boldsymbol{r}_i$$

is a function of position coordinates only. Its value in the state $\Psi$ is therefore

$$\langle \boldsymbol{\mu} \rangle = e \sum_{\alpha=1}^{\nu} Z_\alpha \boldsymbol{R}_\alpha - e \int \Psi^* \left( \sum_{i=1}^{N} \boldsymbol{r}_i \right) \Psi \, d\tau$$

(the nuclear component of $\boldsymbol{\mu}$ is a constant in the Born–Oppenheimer approximation).

The most general form of an $N$-electron operator is

$$\mathcal{F} = F_0 + \sum_{i=1}^{N} \mathcal{F}_1(i) + \sum_{i>j=1}^{N} \mathcal{F}_2(i,j) + ... + \mathcal{F}_N(1, 2, ..., N)$$

where $F_0$ is a constant, $\mathscr{F}_1(i)$ is a one-electron operator involving in general the coordinates and momentum operators of electron $i$ only, $\mathscr{F}_2(i,j)$ is a two-electron operator, and so on. We see that whereas the Hamiltonian $\mathscr{H}$ contains both one- and two-electron terms, the dipole moment is essentially a one-electron property. Other one-electron properties are the quadrupole and other electric multipole moments, the nucleus–electron interaction energy, the force exerted on a nucleus by the electrons and the electric field gradient at a nucleus. In addition, the classical expressions for all these properties are functions of the position coordinates only of the electrons and do not involve their momenta. The corresponding operators are therefore simply the classical functions, and have the general form

$$\mathscr{F} = F(\boldsymbol{r}_1, \boldsymbol{r}_2, ..., \boldsymbol{r}_N) = F_0 + \sum_{i=1}^{N} F_1(\boldsymbol{r}_i) \tag{6.12}$$

The value of such a one-electron property is determined completely by the charge density $P(\boldsymbol{r})$:

$$\langle F \rangle = \int \Psi^* \mathscr{F} \Psi \, d\tau = F_0 + \sum_{i=1}^{N} \int \Psi^* F_1(\boldsymbol{r}_i) \Psi \, d\tau$$

$$= F_0 + \int F_1(\boldsymbol{r}) \rho(\boldsymbol{r}, \sigma) \, dv \, d\sigma$$

$$= F_0 + \int F_1(\boldsymbol{r}) P(\boldsymbol{r}) \, dv \tag{6.13}$$

This expression has a simple classical interpretation. Once the charge density $P(\boldsymbol{r})$ has been calculated, it can be treated as a classical charge distribution function, and the expectation value $\langle F \rangle$ is then simply the classical value of the property $F$ in the system of static charges. For example, the classical dipole moment of a system consisting of $\nu$ charges $Z_\alpha e$ at positions $\boldsymbol{R}_\alpha$ and a distribution of negative charge with density $-eP(\boldsymbol{r})$ is

$$\boldsymbol{\mu} = e \sum_{\alpha=1}^{\nu} Z_\alpha \boldsymbol{R}_\alpha - e \int \boldsymbol{r} P(\boldsymbol{r}) \, dv$$

and this is identical to the quantum-mechanical expectation value $\langle \boldsymbol{\mu} \rangle$.

### 6.2.1 *Intramolecular forces*

The potential-energy function (6.11) of a molecular system of $\nu$ nuclei and $N$ electrons can be written as

$$V = V_\alpha + V' \tag{6.14}$$

## The electron distribution

where

$$V_\alpha = \frac{Z_\alpha e^2}{4\pi\epsilon_0}\left[-\sum_{i=1}^{N}\frac{1}{r_{i\alpha}} + \sum_{\substack{\beta=1 \\ (\beta\neq\alpha)}}^{\nu}\frac{Z_\beta}{R_{\alpha\beta}}\right]$$

is the (classical) potential energy of nucleus $\alpha$ in the presence of the electrons and the other nuclei, and $V'$ is the potential energy of the system of charges in the absence of nucleus $\alpha$. The electrostatic force acting on nucleus $\alpha$ is

$$\boldsymbol{F}_\alpha = \frac{Z_\alpha e^2}{4\pi\epsilon_0}\left[-\sum_i\frac{\boldsymbol{r}_{i\alpha}}{r_{i\alpha}^3} + \sum_{\beta(\neq\alpha)}\frac{Z_\beta \boldsymbol{R}_{\beta\alpha}}{R_{\alpha\beta}^3}\right]$$

where $\boldsymbol{r}_{i\alpha} = \boldsymbol{R}_\alpha - \boldsymbol{r}_i$ and $\boldsymbol{R}_{\beta\alpha} = \boldsymbol{R}_\alpha - \boldsymbol{R}_\beta$. In particular, the component of the force along the $x$ direction is

$$F_{x\alpha} = -\frac{\partial V_\alpha}{\partial X_\alpha} = \frac{Z_\alpha e^2}{4\pi\epsilon_0}\left[-\sum_i\frac{(X_\alpha - x_i)}{r_{i\alpha}^3} + \sum_{\beta(\neq\alpha)}\frac{Z_\beta(X_\alpha - X_\beta)}{R_{\alpha\beta}^3}\right]$$

where, for example, $X_\alpha$ and $x_i$ are the $x$ coordinates of nucleus $\alpha$ and electron $i$ respectively. In addition, since $V'$ in (6.14) is independent of $\boldsymbol{R}_\alpha$, the force on nucleus $\alpha$ is also given by the derivative of the total potential energy:

$$F_{x\alpha} = -\frac{\partial V}{\partial X_\alpha}$$

The force $\boldsymbol{F}_\alpha$ is a one-electron quantity of the type (6.12), and its expectation value in the electronic state $\Psi$ is

$$\langle \boldsymbol{F}_\alpha\rangle = \int \Psi^* \boldsymbol{F}_\alpha \Psi\, d\tau$$

$$= \frac{Z_\alpha e^2}{4\pi\epsilon_0}\left[\sum_{\beta\neq\alpha}\frac{Z_\beta \boldsymbol{R}_{\beta\alpha}}{R_{\alpha\beta}^3} - \int\frac{(\boldsymbol{R}_\alpha - \boldsymbol{r})}{|\boldsymbol{R}_\alpha - \boldsymbol{r}|^3}P(\boldsymbol{r})\,dv\right] \qquad (6.15)$$

and may be identified with the value of the force on nucleus $\alpha$ in this state. The forces acting on the nuclei of a molecule can then be given the classical interpretation discussed above.

*The Hellmann–Feynman theorem.* The identification of the expectation value $\langle \boldsymbol{F}_\alpha\rangle$ with the force on nucleus $\alpha$ is a consequence of the Hellmann–Feynman theorem. Quite generally, let $\mathscr{H}$ be the Hamiltonian of an $N$-electron system and let the potential energy $V$ (but not the kinetic-energy operator) depend on one or more parameters $(\lambda_1, \lambda_2, \ldots)$ which are independent of the coordinates of the electrons:

$$\mathscr{H} = \frac{-\hbar^2}{2m_e}\sum_{i=1}^{N}\nabla_i^2 + V(\boldsymbol{r}_1, \boldsymbol{r}_2, \ldots, \boldsymbol{r}_N; \lambda_1, \lambda_2, \ldots)$$

Such a parameter may be a nuclear coordinate or bond length in a molecule, or it may be the strength of an externally applied field. The eigenfunctions and eigenvalues of $\mathcal{H}$ also depend on the values of these parameters, and the variation of the energy with respect to variations of such a parameter, $\lambda$ say, is given by

$$\frac{\partial E}{\partial \lambda} = \frac{\partial}{\partial \lambda} \int \Psi^* \mathcal{H} \Psi \, d\tau$$

$$= \int \Psi^* \left(\frac{\partial \mathcal{H}}{\partial \lambda}\right) \Psi \, d\tau + \int \Psi^* \mathcal{H} \left(\frac{\partial \Psi}{\partial \lambda}\right) d\tau + \int \left(\frac{\partial \Psi^*}{\partial \lambda}\right) \mathcal{H} \Psi \, d\tau \quad (6.16)$$

The last two terms of (6.16), involving the derivative of the wave function, can be shown to vanish when $\Psi$ is an exact eigenfunction of $\mathcal{H}$. In addition, because only $V$ in $\mathcal{H}$ has been assumed to depend on $\lambda$, $(\partial \mathcal{H}/\partial \lambda)$ can be replaced by $(\partial V/\partial \lambda)$, and (6.16) reduces to

$$\frac{\partial E}{\partial \lambda} = \int \Psi^* \left(\frac{\partial V}{\partial \lambda}\right) \Psi \, d\tau \quad (6.17)$$

This is the Hellmann–Feynmann theorem. In particular, when $\lambda$ is a geometric parameter, for example the $x$ coordinate $X_\alpha$ of nucleus $\alpha$, it follows from (6.17) that

$$-\frac{\partial E}{\partial X_\alpha} = \int \Psi^* \left(-\frac{\partial V}{\partial X_\alpha}\right) \Psi \, d\tau$$

which is the $x$ component $\langle F_{x\alpha} \rangle$ of the expectation value (6.15). The identification of $\langle F_\alpha \rangle$ with the force on nucleus $\alpha$ then follows from the interpretation of the energy $E$, as a function of the nuclear coordinates, as the potential-energy function for the motion of the nuclei in the molecule (in the Born–Oppenheimer approximation). This interpretation implies that $(-\partial E/\partial X_\alpha)$, and therefore $\langle F_{x\alpha} \rangle$, is the $x$ component of the force on nucleus $\alpha$.

The Hellmann–Feynman theorem is valid for exact eigenfunctions of the Hamiltonian and for certain classes of approximate wave functions, including accurate Hartree–Fock wave functions. It is not in general valid for an arbitrary wave function, such as an approximate SCF function, and the theorem must then be used with some caution when invoked in a discussion of, say, intramolecular forces.

## 6.3 CHARGE-DENSITY MAPS

A visual picture of the shapes and sizes of molecules, and of the relative tightness of binding in different molecules, is conveniently obtained from contour maps of the total charge density $P(r)$. Density maps (drawn

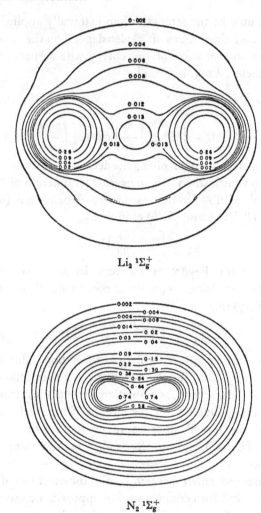

Fig. 6.2. Total molecular charge-density maps for the ground states of $Li_2$ and $N_2$. The values of the contours are in the (atomic) unit of inverse volume $a_\infty^{-3}$. The innermost, circular, contours centred on the nuclei have been omitted for clarity. The values of the density at the nucleus are $13.9a_\infty^{-3}$ for $Li_2$ and $205.6a_\infty^{-3}$ for $N_2$. The two maps are drawn to the same scale (Bader, Henneker and Cade 1967).

to the same scale) for the ground states of $Li_2$ and $N_2$ are shown in fig. 6.2 (Bader, Henneker and Cade 1967). The contours lie in a plane which contains the nuclei, and the outermost (0.002) contour, which includes over 95 per cent of the electronic charge, provides a measure of the dimensions of the molecules. We see that although the bond length in $Li_2$ is more than twice that in $N_2$, the effective lengths of the two molecules are not very different. This can be explained within the orbital approximation by the fact that the lithium atom, unlike nitrogen, has only one valence electron, and the removal of this electron from the vicinity of the nucleus into the bonding region between the nuclei of the molecule leaves essentially only the tightly bound inner-shell density around each nucleus.

The redistribution of charge that accompanies the formation of a chemical bond may be described in terms of the density difference

$$\Delta P(\mathbf{r}) = P_{mol}(\mathbf{r}) - \sum_{atoms} P_{atom}(\mathbf{r})$$

which is obtained by subtracting the densities of the unperturbed constituent atoms, placed at the appropriate positions, from the molecular density. The density-difference maps corresponding to fig. 6.2 for $Li_2$ and $N_2$ are shown in fig. 6.3. The map for $N_2$ in particular shows not only the expected build-up of charge between the nuclei but also in the regions behind the nuclei (away from the bonding region), with a corresponding depletion of charge close to the nuclei and in the outer regions of the molecule. Any interpretation of the nature of the covalent bond must explain how this redistribution of charge is necessary for and gives rise to a stabilization of the molecule relative to the separate atoms. The interpretation of the nature of the chemical bond is discussed in greater detail in chapter 7.

The overall charge distribution in a molecule may be analysed in terms of the amounts of charge found in different regions of the space, and the forces exerted on the nuclei by the charge in these different regions. The simplest partitioning of the molecular space is into 'binding' and 'antibinding' regions (Berlin 1951). These regions are defined such that, in a diatomic molecule for example, the forces exerted on the nuclei by an electron in the binding region tends to pull the nuclei together, whilst an electron in the antibinding region tends to pull the nuclei apart. The boundaries between binding and antibinding regions are shown by the dotted curves in fig. 6.3, with the binding region lying between the nuclei, and the nuclei on the boundaries. The $\Delta P$ maps for $Li_2$ and $N_2$ show that the original atomic densities have been distorted so as to move charge into the antibinding regions as well as into the binding region. Of the first-row homonuclear diatomic molecules, $Li_2$ to $F_2$, only $Li_2$ fits the picture of a simple transfer of charge into the binding region, with very little

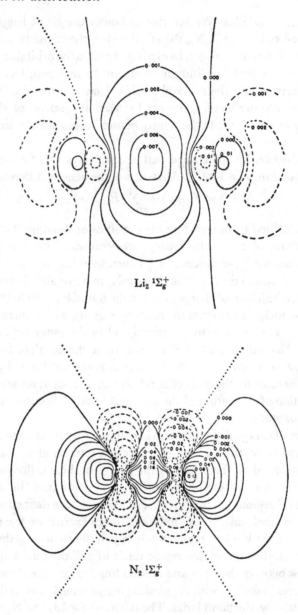

Fig. 6.3. Density-difference maps for $Li_2$ and $N_2$. The solid lines are positive contours, the dashed lines are negative contours, and the dotted lines (shown in full for $N_2$) separate the binding from the antibinding regions (Bader, Henneker and Cade 1967).

transfer into the antibinding regions. This, and the diffuse nature of the $\Delta P$ distribution, is characteristic of bonds which, in the LCAO–MO approximation, can be described by the overlap of essentially spherical atomic orbitals with little directional atomic $p$-type character. The two-way transfer of charge in $N_2$, with a strong concentration near the centre of the molecule, is typical of an increased participation of $p_\sigma$ atomic orbitals. On going from $Li_2$ across the series to $F_2$, charge is found to be increasingly removed into the antibinding regions. There is apparently no simple correlation between the net increase of charge in the binding region on bond formation and the strength of the bond, and at least as important as the total charges in the two regions is the way the charge is distributed within the regions. Thus the distribution in $Li_2$ is very diffuse in the binding region, whilst that in $N_2$ is concentrated about the bond axis where it can exert a maximum force of attraction on the nuclei, in contrast to the antibinding regions in which the distribution is more diffuse.

A study of the density maps for heteronuclear diatomic molecules sheds light on the nature of ionic bonding as compared with covalent bonding. Whereas a covalent bond is one for which the $\Delta P$ map shows a density increase in the region between the nuclei which is shared about equally by both nuclei, the $\Delta P$ map for LiF (fig. 6.4) shows that the

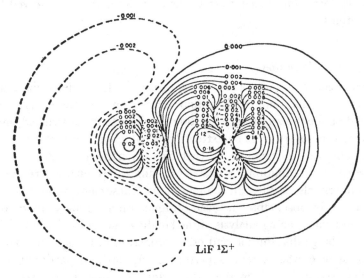

Fig. 6.4. Density-difference map for LiF. The solid lines are positive contours, the dashed lines are negative contours (Bader, Henneker and Cade 1967).

increase in charge density which binds the nuclei is localized almost entirely on the fluorine. In this extreme example of ionic bonding, the formation of the bond can be described in terms of the complete transfer of an electron from the lithium onto the fluorine, with a subsequent back-polarization towards the lithium (Bader and Henneker 1965). The remaining charge on the lithium is also polarized away from the fluorine, presumably as a result of the repulsive field of the anion. It may be noted also that the transfer of charge from the lithium onto the fluorine is nevertheless accompanied by a decrease in charge close to the fluorine nucleus. This decrease appears to be an important requirement for the formation of a stable bond, be it covalent or ionic. Similar studies of the first-row diatomic hydrides (Bader, Keaveny and Cade 1967) show that the bonding in LiH is best classified as ionic, with a $\Delta P$ distribution somewhat similar in type to that in LiF. The molecules BH to HF are essentially covalent, with a shared density increase between the nuclei which is increasingly polarized away from the hydrogen, whilst the bonding in BeH, and to some extent in LiH, is transitional between ionic and covalent. The second-row hydrides (Cade *et al.* 1969) show a similar trend, with NaH ionic and with MgH and AlH transitional. The density maps for the hydrides also suggest that hydrogen is anomalous amongst atoms in that the charge density at the nucleus increases on bond formation, in contrast to the decrease normally observed for other atoms.

**6.3.1** *Orbital densities*

Although concepts such as resonance, directed valence, delocalized and localized orbitals, and configuration interaction are merely artifacts which describe the way a wave function may have been constructed, such concepts can nevertheless be extremely valuable for interpretative purposes. They can, for example, be used to partition the charge distribution into separate contributions associated with different resonance structures, different orbitals, or different configurations. These contributions, whilst not normally observable or even unique, often allow a more detailed and enlightening analysis of the bonding in a molecule, and of the way the charge distribution and other properties of the system change as the state of the system is perturbed. An obvious partitioning of the charge distribution in the orbital approximation is into separate orbital contributions, in accordance with (6.7). Thus, for a single determinant containing a number of singly occupied and doubly occupied orbitals,

$$P(\boldsymbol{r}) = \sum_i n_i |\psi_i(\boldsymbol{r})|^2 = \sum_i n_i P_i(\boldsymbol{r}) \qquad (6.18)$$

where $P_i(\boldsymbol{r}) = |\psi_i(\boldsymbol{r})|^2$ is the charge-density function of orbital $\psi_i$, and $n_i = 1$ or $2$ is the occupation number of the orbital.

In fig. 6.5 are shown contour maps of the valence molecular orbitals (from which the orbital densities are readily obtained) for the water molecule in its ground-state orbital configuration $(1a_1)^2(2a_1)^2(1b_2)^2(3a_1)^2$ $(1b_1)^2$ (Dunning, Pitzer and Aung 1972). The contours for $2a_1$, $1b_2$ and $3a_1$ lie in the principal molecular plane, and for $1b_1$ in the bisecting plane perpendicular to the principal plane. These orbitals have been obtained from an accurate SCF calculation involving a large basis of Gaussian functions, including $d$-type polarization functions on the oxygen and $p$-type polarization functions on the hydrogens. The molecular orbitals obtained from a somewhat simpler calculation involving only a minimal Slater-function basis (Aung, Pitzer and Chan 1968) are shown in table 6.1. The $1a_1$ molecular orbital is essentially the oxygen $1s$ atomic orbital, with small contributions from other atomic orbitals on all three atoms. $1b_2$ is a straightforward bonding orbital, and $1b_1$ a lone-pair orbital on the oxygen. The interpretation of the orbitals $2a_1$ and $3a_1$ is not so obvious however. The $2a_1$ molecular orbital is mainly the oxygen $2s$ atomic orbital, but with significant contributions from the oxygen $2p_z$ and the hydrogen $1s$. The $3a_1$ molecular orbital contains appreciable contributions from both the oxygen $2s$ and $2p_z$, as well as from the hydrogen $1s$. The $2a_1$ orbital is usually regarded as a bonding orbital, and $3a_1$ as a weakly bonding lone-pair orbital (for example, Herzberg 1967), on the grounds that the hybridization of the oxygen $2s$ and $2p_z$ atomic orbitals in $3a_1$ concentrates electronic charge on the side of the oxygen away from the hydrogens. This interpretation is only partially confirmed by the

TABLE 6.1  *Molecular-orbital coefficients for the water molecule (Aung, Pitzer and Chan 1968)*

| MO | Oxygen AO | | | | | Hydrogen AO | |
|---|---|---|---|---|---|---|---|
| | $1s$ | $2s$ | $2p_z$ | $2p_x$ | $2p_y$ | $1s_1$ | $1s_2$ |
| $1a_1$ | 0.997 | 0.015 | 0.003 | 0 | 0 | −0.004 | −0.004 |
| $2a_1$ | −0.222 | 0.843 | 0.132 | 0 | 0 | 0.151 | 0.151 |
| $3a_1$ | −0.093 | 0.516 | −0.787 | 0 | 0 | −0.264 | −0.264 |
| $1b_2$ | 0 | 0 | 0 | 0.624 | 0 | 0.424 | −0.424 |
| $1b_1$ | 0 | 0 | 0 | 0 | 1 | 0 | 0 |

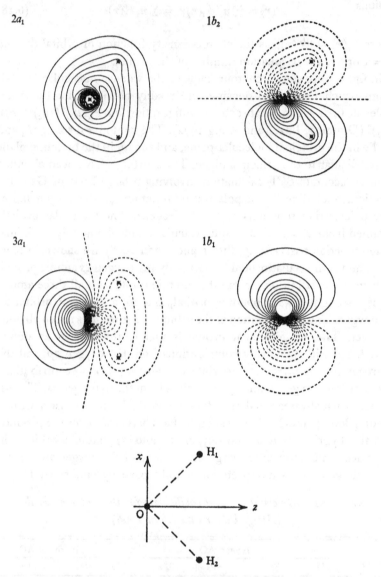

Fig. 6.5. Contour maps of the valence molecular orbitals in the ground state of the water molecule. The contours for $2a_1$, $1b_2$ and $3a_1$ lie in the principal molecular plane ($xz$), and for $1b_1$ in the bisecting plane ($yz$) perpendicular to the principal plane. The solid lines are positive contours, the dashed lines are negative contours (Dunning, Pitzer and Aung 1972).

corresponding orbital maps in fig. 6.5. It has been argued by some authors on the other hand (for example, Murrell, Kettle and Tedder 1970) that it is the $2a_1$ orbital which should be regarded as non-bonding, since it is essentially the oxygen $2s$ atomic orbital. These conflicting interpretations demonstrate the dangers inherent in such qualitative arguments, and we shall see in §6.4.3 how the conflict may be resolved in the present case by a transformation of the molecular orbitals to localized orbitals.

## 6.4 LOCALIZED ORBITALS

### 6.4.1 *Localized and delocalized orbitals*

The molecular orbitals in the restricted Hartree–Fock model are symmetry orbitals which belong to the irreducible representations of the symmetry group of the molecule, and they are usually delocalized orbitals extending over several or all the nuclei of the molecule. For example, the ground state of $CH_4$ has the orbital configuration $(1a_1)^2 (2a_1)^2 (1t_2)^6$, and the RHF wave function is the Slater determinant

$$\Psi_{MO} = |1a_1\alpha, 1a_1\beta, 2a_1\alpha, 2a_1\beta, 1t_2\alpha, 1t_2\beta, 1t_2'\alpha, 1t_2'\beta, 1t_2''\alpha, 1t_2''\beta|$$

$$(6.19)$$

The molecular orbital $1a_1$ is the $1s$ atomic orbital on the carbon with only small contributions from other atomic orbitals on all five atoms. The orbital $2a_1$, on the other hand, is delocalized over all five nuclei, and is essentially a mixture of the $2s$ atomic orbital on the carbon and a symmetrical combination of the $1s$ orbitals on the four hydrogens. The three $1t_2$ orbitals are made up of $p$-type orbitals on the carbon and suitable combinations of atomic orbitals on the hydrogens. This description of the electronic structure of the methane molecule is closedly related to that of the neon atom, for which the ground-state orbital configuration is $(1s)^2 (2s)^2 (2p)^6$. It apparently bears little relation however, to the conventional picture of methane, in which the bonding is interpreted in terms of four equivalent bond orbitals which are localized about the C–H axes. A wave function which gives such a description has the form

$$\Psi_{LOC} = |1s_C\,\alpha, 1s_C\beta, b_1\alpha, b_1\beta, b_2\alpha, b_2\beta, b_3\alpha, b_3\beta, b_4\alpha, b_4\beta| \quad (6.20)$$

where $1s_C$ is the $1s$ atomic orbital on the carbon and, for example, $b_1$ is a bond orbital which, in a simple treatment, might be represented by a linear combination of an $sp^3$ hybrid orbital on the carbon and a $1s$ orbital on that hydrogen towards which the carbon hybrid is directed.

## The electron distribution

These two apparently exclusive descriptions of the system can be reconciled by noting that a Slater determinant made up of doubly occupied orbitals is invariant with respect to certain linear transformations of the orbitals (see §3.1). Given a normalized $N$-electron single-determinant wave function constructed from $N/2$ doubly occupied orthonormal orbitals,

$$\Psi = |\psi_1\alpha, \psi_1\beta, ..., \psi_{N/2}\alpha, \psi_{N/2}\beta|$$

a linear transformation of the orbitals,

$$\lambda_k = \sum_{n=1}^{N/2} \psi_n C_{nk} \quad (k = 1, 2, ..., N/2) \tag{6.21}$$

gives a new wave function

$$\Psi' = |\lambda_1\alpha, \lambda_1\beta, ..., \lambda_{N/2}\alpha, \lambda_{N/2}\beta|$$

which is identical to the original function $\Psi$ if the transformation conserves orthonormality; that is, if the new orbitals $\lambda_k$ are also normalized and orthogonal. Such a transformation is called *unitary*, and the coefficients in (6.21) satisfy the relations

$$\sum_{n=1}^{N/2} C_{nk}^* C_{nl} = \delta_{kl}, \quad \sum_{k=1}^{N/2} C_{mk}^* C_{nk} = \delta_{mn}$$

The two sets of orbitals provide *alternative* but *equivalent* pictures of the electronic structure in the same state of the system, and since there is an infinite number of possible unitary transformations of the type (6.21), there is a corresponding infinite number of possible equivalent pictures. Because the wave function is invariant, so are also all derived properties and, in particular, the charge distribution:

$$P(\mathbf{r}) = 2 \sum_{n=1}^{N/2} |\psi_n(\mathbf{r})|^2 = 2 \sum_{k=1}^{N/2} |\lambda_k(\mathbf{r})|^2$$

This property of a single-determinant wave function allows the orbitals to be chosen in the most convenient way for the interpretation of any particular property of the system. Thus, the delocalized molecular orbitals are most suitable for a discussion of the electronic spectrum. For example, an approximate value of the first ionization energy of the methane molecule is given by the orbital energy of a $1t_2$ orbital, and an approximate wave function for the ion (assuming a tetrahedral geometry) is obtained from (6.19) by the removal of an electron from one of the $1t_2$ orbitals. The corresponding description in terms of localized orbitals requires a wave function which is a linear combination of four Slater determinants, each of which is obtained from (6.20) by the removal of an electron from a bond orbital. On the other hand, properties such as the energy of formation, dipole moment and stereochemistry are often interpreted most conveniently in terms of localized bond and lone-pair orbitals, and the properties of related molecules may be correlated in terms of the approximate transferability of localized orbitals from one molecule to another.

### 6.4.2 *Localization procedures*

Several methods have been proposed for the localization of molecular (and atomic) orbitals (for example: Boys 1960; Edmiston and Ruedenberg 1963; Magnasco and Perico 1967; von Niessen 1972), and all give rather similar localized orbitals. The most general and non-arbitrary method (Edmiston and Ruedenberg 1963) is based on the minimization of inter-orbital interactions. We consider a closed-shell $N$-electron wave function, for which the energy (3.25) is

$$E = 2 \sum_{n=1}^{N/2} f_n + \sum_{\alpha>\beta=1}^{\nu}\sum \frac{Z_\alpha Z_\beta}{R_{\alpha\beta}} + \sum_{n=1}^{N/2} J_{nn} + \sum_{\substack{m,\,n=1\\(m\neq n)}}^{N/2}\sum (2J_{mn} - K_{mn})$$

The quantity $J_{nn}$ is the self-energy of a pair of electrons in the orbital $\psi_n$, and is a maximum when the electrons are on average as close to each other as possible. The maximization of the total self-energy,

$$D = \sum_{n=1}^{N/2} J_{nn}$$

therefore provides a criterion for the localization of the electrons in pairs. It can be shown that the quantities

$$C = 2\sum_{m,\,n}\sum J_{mn} \quad \text{and} \quad X = \sum_{m,\,n}\sum K_{mn}$$

are, like the total energy $E$, separately invariant with respect to a unitary transformation of the orbitals (6.21). The maximization of the total self-energy $D$ is therefore equivalent to the minimization of the total inter-orbital Coulomb energy

$$C - 2D = 2\sum_{m\neq n}\sum J_{mn}$$

In addition, since $K_{nn} = J_{nn}$, the maximization of $D$ is also equivalent to

*The electron distribution*

the minimization of the total exchange energy

$$X - D = \sum_{m \neq n} \sum K_{mn}$$

The condition that the total self-energy be a maximum may be used as a criterion for localization even when the wave function is not a single Slater determinant of doubly occupied orbitals. In addition, a subclass of orbitals may be localized separately if the summation in $D$ is restricted to the orbitals of the subclass.

As an example of the localization of orbitals, we consider the pair of atomic orbitals $\psi_1 = 2s$ and $\psi_2 = 2p_z$ on the same centre, and the linear combinations

$$\lambda_1 = C\psi_1 + S\psi_2, \quad \lambda_2 = C\psi_2 - S\psi_1$$

The new orbitals, $\lambda_1$ and $\lambda_2$, are orthonormal if the coefficients satisfy the relation $C^2 + S^2 = 1$. The coefficients may therefore be expressed as the sine and cosine of some angle $\theta$, whose value defines the transformation. We assume, for simplicity, that the orbitals are real in general. Then

$$D = (\lambda_1^2 | \lambda_1^2) + (\lambda_2^2 | \lambda_2^2)$$

where, in general for real orbitals,

$$(\lambda_i \lambda_j | \lambda_k \lambda_l) = \int\int \frac{[\lambda_i(r_1)\lambda_j(r_1)][\lambda_k(r_2)\lambda_l(r_2)]}{r_{12}} \, dv_1 \, dv_2$$

In this notation, $J_{ij} = (\lambda_i^2 | \lambda_j^2)$ and $K_{ij} = (\lambda_i \lambda_j | \lambda_i \lambda_j)$ are the Coulomb and exchange terms of a pair of electrons in orbitals $\lambda_i$ and $\lambda_j$. In the present case, if $C = \cos\theta$ and $S = \sin\theta$, $D$ is a function of $\theta$, and its maximum, or minimum, value is determined by the condition $\partial D/\partial\theta = 0$. The corresponding orbitals satisfy

$$(\lambda_1^2 | \lambda_1 \lambda_2) = (\lambda_2^2 | \lambda_1 \lambda_2) \tag{6.22}$$

and this equation can be solved for $\theta$. In terms of the original orbitals, $2s$ and $2p_z$, (6.22) reduces to

$$(C^2 - S^2) \, CS(J_{2s,\,2s} + J_{2p_z,\,2p_z} - 2J_{2s,\,2p_z} - 4K_{2s,\,2p_z}) = 0$$

The independent solutions are (i) $C = 1$, $S = 0$, and (ii) $C^2 = S^2 = \frac{1}{2}$. The first of these corresponds to the minimum value of $D$, and shows that the original orbitals are the least localized. The most localized orbitals are the equivalent digonal $sp$ hybrids $(2s \pm 2p_z)/\sqrt{2}$. In the same way, for example, the localized orbitals obtained from the atomic orbitals $2s$, $2p_x$, $2p_y$ and $2p_z$ are the equivalent tetrahedral $sp^3$ hybrids. In general, for real orbitals, $D$ has its maximum value when the orbitals satisfy the conditions

$$(\lambda_i^2 | \lambda_i \lambda_j) = (\lambda_j^2 | \lambda_i \lambda_j) \tag{6.23}$$

for all pairs of orbitals.

What are conventionally known as the molecular orbitals of an $N$-electron molecule are the eigenfunctions of the corresponding Hartree–Fock operator; for a closed-shell state in the RHF approximation,

$$h^{\mathrm{RHF}}\psi_n = \epsilon_n \psi_n \quad (n = 1, 2, ..., N/2) \tag{3.26}$$

A unitary transformation (6.21) to new orbitals $\lambda_k$ leaves the Hartree–Fock operator $h^{\mathrm{RHF}}$ unchanged, and the new orbitals are solutions of the modified (and more general) Hartree–Fock equations

$$h^{\mathrm{RHF}}\lambda_k = \sum_{l=1}^{N/2} \lambda_l \epsilon_{lk} \quad (k = 1, 2, ..., N/2) \tag{6.24}$$

where 
$$\epsilon_{lk} = \int \lambda_l^* h^{\mathrm{RHF}} \lambda_k \, dv$$

The solutions of (6.24) are unique only within a unitary transformation of the orbitals. The particular solutions for which the matrix $\epsilon$ has the diagonal form, $\epsilon_{nn} = \epsilon_n$ and $\epsilon_{mn} = 0$ if $m \neq n$, are often called the *canonical* molecular orbitals. These are the eigenfunctions of $h^{\mathrm{RHF}}$ and they have a special significance only in that the eigenvalues $\epsilon_n$ can be interpreted as orbital energies. We shall continue to call them simply the molecular orbitals.

The localized orbitals obtained by the above method of Edmiston and Ruedenberg, and by most other localization methods, are usually in broad agreement with the conventional picture of the charge distribution and bonding in a molecule in terms of lone-pair and bond orbitals. Thus, for example, $CH_4$ has a localized inner-shell orbital on the carbon which is essentially the carbon $1s$ atomic orbital, with very small contributions from other atomic orbitals on all five atoms, and four equivalent two-centre bond orbitals localized about the bond axes, again with small contributions from other atomic orbitals. A typical C–H bond orbital is illustrated in fig. 6.6.

### 6.4.3 *The ground state of* $H_2O$

In table 6.2 are summarized the results of an Edmiston–Ruedenberg localization of a set of molecular orbitals for $H_2O$ almost identical to those given in table 6.1 (Edmiston and Ruedenberg 1966). The inner-shell lone-pair orbital $i$ is almost identical to the (canonical) molecular orbital $1a_1$, and is essentially the oxygen $1s$ atomic orbital. The

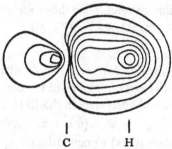

C    H

Fig. 6.6. The C–H bond orbital (Rothenberg 1969).

### TABLE 6.2 *Localized orbitals in the water molecule*
### *(Edmiston and Ruedenberg 1966)*

Coefficients for the transformation of canonical molecular orbitals to localized orbitals

|           | $1a_1$  | $2a_1$  | $3a_1$  | $1b_2$  | $1b_1$  |
| --------- | ------- | ------- | ------- | ------- | ------- |
| $i$       | 0.990   | 0.127   | 0.063   | 0       | 0       |
| $l_1$     | −0.090  | 0.424   | 0.559   | 0       | 0.707   |
| $l_2$     | −0.090  | 0.424   | 0.559   | 0       | −0.707  |
| $bOH_1$   | −0.044  | 0.559   | −0.432  | 0.707   | 0       |
| $bOH_2$   | −0.044  | 0.559   | −0.432  | −0.707  | 0       |

LCAO expansion coefficients of the localized orbitals

|         | Oxygen AO |         |         |         |         | Hydrogen AO |         |
| ------- | --------- | ------- | ------- | ------- | ------- | ----------- | ------- |
|         | $1s$      | $2s$    | $2p_z$  | $2p_x$  | $2p_y$  | $1s_1$      | $1s_2$  |
| $i$     | 0.992     | −0.120  | 0.035   | 0       | 0       | −0.004      | −0.004  |
| $l_1$   | 0.095     | 0.653   | −0.376  | 0       | 0.707   | −0.102      | −0.102  |
| $l_2$   | 0.095     | 0.653   | −0.376  | 0       | −0.707  | −0.102      | −0.102  |
| $bOH_1$ | 0.014     | 0.226   | 0.414   | 0.412   | 0       | 0.566       | −0.160  |
| $bOH_2$ | 0.014     | 0.226   | 0.414   | −0.412  | 0       | −0.160      | 0.566   |

transformation from the valence-shell molecular orbitals to localized orbitals is best considered in three steps.

(i) Localization of $2a_1$ and $3a_1$,

$$la_1 = 0.6(2a_1) + 0.8(3a_1)$$
$$ba_1 = 0.8(2a_1) - 0.6(3a_1)$$

(6.25)

(ii) Localization of $ba_1$ and $1b_2$,

$$bOH_1 = (1/\sqrt{2})(ba_1 + 1b_2)$$
$$bOH_2 = (1/\sqrt{2})(ba_1 - 1b_2)$$

(6.26)

(iii) Localization of $la_1$ and $1b_1$,

$$l_1 = (1/\sqrt{2})(la_1 + 1b_1)$$
$$l_2 = (1/\sqrt{2})(la_1 - 1b_1)$$

(6.27)

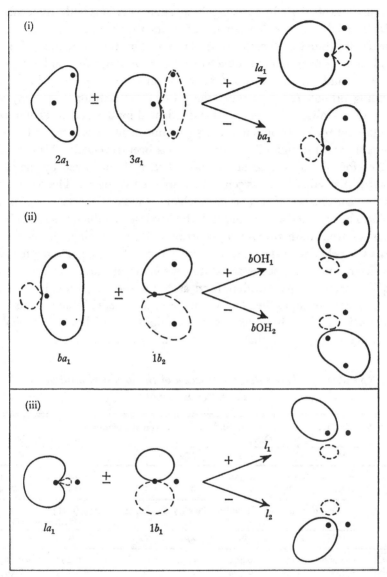

Fig. 6.7. Schematic representation of the three-step localization of the orbitals in the ground state of $H_2O$. Diagrams (i) and (ii) lie in the principal molecular plane, and (iii) in the bisecting plane perpendicular to the principal plane.

## The electron distribution

The orbitals obtained by applying these transformations to the molecular orbitals in table 6.1 are shown in table 6.3, and the three-step transformation is illustrated schematically in fig. 6.7. The transformation (6.25) produces a pair of totally symmetric orbitals, of which $la_1$ can be interpreted unambiguously as a non-bonding lone-pair orbital and $ba_1$ as a bonding orbital. $la_1$, containing the greater contribution from $3a_1$, is somewhat similar to $2a_1$ in that it is dominated by the oxygen $2s$ atomic orbital, but the contribution of the $2p_z$ atomic orbital has increased considerably and the sign of its coefficient has been reversed. This orbital is therefore the oxygen $2s$ atomic orbital strongly polarized by the $2p_z$ towards that side of the oxygen away from the hydrogens. The problem discussed at the end of §6.3, as to which of the molecular orbitals $2a_1$ and $3a_1$ is to be regarded as bonding and which as non-bonding, has therefore been resolved. Both $2a_1$ and $3a_1$ contribute to the bonding, and neither can be labelled as essentially bonding or non-bonding, although the coefficients in (6.25) suggest that $2a_1$ is somewhat more bonding than $3a_1$. The argument used to identify $3a_1$ as the non-bonding orbital and that used to identify $2a_1$ as the non-bonding orbital can *both* be applied to $la_1$ and *neither* to $ba_1$ which, like $1b_2$ is a straightforward bonding orbital. This analysis suggests that the (canonical) molecular orbitals are not

TABLE 6.3    *Three-step construction of the localized orbitals for the water molecule*

| | Oxygen AO | | | | | Hydrogen AO | |
|---|---|---|---|---|---|---|---|
| | $1s$ | $2s$ | $2p_z$ | $2p_x$ | $2p_y$ | $1s_1$ | $1s_2$ |
| | (i) Localization of $2a_1$ and $3a_1$ by the transformation (6.25) | | | | | | |
| $la_1$ | −0.208 | 0.919 | −0.550 | o | o | −0.121 | −0.121 |
| $ba_1$ | −0.122 | 0.365 | 0.578 | o | o | 0.279 | 0.279 |
| | (ii) Construction of bond orbitals by the transformation (6.26) | | | | | | |
| | $1s$ | $2s$ | $2p_z$ | $2p_x$ | $2p_y$ | $1s_1$ | $1s_2$ |
| $bOH_1$ | −0.086 | 0.258 | 0.409 | 0.441 | o | 0.497 | −0.103 |
| $bOH_2$ | −0.086 | 0.258 | 0.409 | −0.441 | o | −0.103 | 0.497 |
| | (iii) Construction of lone-pair orbitals by the transformation (6.27) | | | | | | |
| | $1s$ | $2s$ | $2p_z$ | $2p_x$ | $2p_y$ | $1s_1$ | $1s_2$ |
| $l_1$ | −0.147 | 0.650 | −0.389 | o | 0.707 | −0.086 | −0.086 |
| $l_2$ | −0.147 | 0.650 | −0.389 | o | −0.707 | −0.086 | −0.086 |

always the most suitable orbitals for a discussion of the bonding in molecules in terms of bonding, non-bonding, and antibonding orbitals.

The second step, (6.26), in the construction of the localized orbitals involves the mixing of the bonding orbitals $ba_1$ and $1b_2$ to give a pair of equivalent two-centre bond orbitals $bOH_1$ and $bOH_2$. The three orbitals $la_1$, $bOH_1$ and $bOH_2$ are the localized orbitals of the $\sigma$ structure of the molecule. The third step involves the mixing of the $\sigma$ and $\pi$ lone-pair orbitals $la_1$ and $1b_1$ to give a pair of equivalent lone-pair orbitals $l_1$ and $l_2$. The final set of four valence localized orbitals differs slightly from that shown in table 6.2 mainly because (*a*) the $1a_1$ molecular orbital has not been included in the transformation, and (*b*) the transformation coefficients in (6.25)–(6.27) do not correspond exactly to those shown in table 6.2. The interpretation is not however altered in any essential way by these small differences.

### 6.4.4 *Multiple bonds*

The occurrence of a multiple bond in a molecule gives rise to localized orbitals rather different from those normally associated with such bonds. Thus, the conventional picture of the electronic structure of the ground state of ethylene involves a pair of inner-shell orbitals on the carbons, four equivalent C–H bond orbitals, and two-centre $\sigma$ and $\pi$ bond orbitals for the double C–C bond. The ground-state orbital configuration of $C_2H_4$ is

$$(1a_g)^2(1b_{1u})^2(2a_g)^2(2b_{1u})^2(1b_{2u})^2(3a_g)^2(1b_{3g})^2(1b_{3u})^2$$

and all the orbitals except $1b_{3u}$ are delocalized $\sigma$ orbitals, whilst $1b_{3u}$ is the two-centre $\pi$ orbital of the C–C bond. The localization of these orbitals can then be considered in two steps, of which the first involves the localization of the $\sigma$ orbitals only. The resulting localized orbitals, together with the $\pi$ orbital, give the conventional picture of the bonding. The second step involves the further localization of the C–C bond orbitals; denoting these by $\sigma_{C-C}$ and $\pi_{C-C}$, the localized C–C bond orbitals are

$$b_1 = (1/\surd 2)(\sigma_{C-C} + \pi_{C-C})$$
$$b_2 = (1/\surd 2)(\sigma_{C-C} - \pi_{C-C})$$

This mixing of $\sigma$ and $\pi$ orbitals is analogous to the mixing (6.27) of the lone-pair $\sigma$ and $\pi$ orbitals in $H_2O$, and is illustrated in fig. 6.8. The resulting orbitals are equivalent 'bent-bond' orbitals, and are the localized orbitals characteristic of multiple bonds (Hall and Lennard–Jones 1951;

# The electron distribution

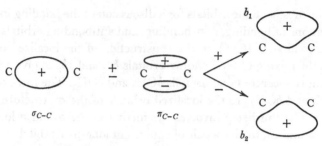

Fig. 6.8. Localization of the double-bond orbitals in $C_2H_4$.

Pople 1957). Similarly, the triple bond in $N_2$ or $C_2H_2$ is represented by a set of three equivalent bond orbitals.

The bent-bond description of the double bond provides a link between the electronic structure of $C_2H_4$ and that of $B_2H_6$ (Switkes, *et al.* 1969). The localized orbitals of diborane include a pair of three-centre bridge orbitals whose shapes resemble the corresponding orbitals in ethylene, as shown in fig. 6.9. The essential difference is that the removal of a proton from each carbon in $C_2H_4$ into the bridge position in $B_2H_6$ results in a shift of the orbital centroids away from the principal molecular plane, so that a direct B–B bond is no longer in evidence. The localized orbitals in other boron hydrides also include two-centre and three-centre orbitals which describe direct bonds between the boron atoms (for example: Switkes, Lipscomb and Newton 1970*b*; Laws, Stevens and Lipscomb 1972; Guest and Hillier 1974). Similar localized orbitals are found in the carboranes (for example: Marynick and Lipscomb 1972; Epstein, Marynick and Lipscomb 1973; Guest and Hillier 1973). In $BeBH_5$, the boron and beryllium atoms appear to be linked by three hydrogen bridges (Ahlrichs 1973), and the electronic structure in this molecule can be related to that in $C_2H_2$, with the triplet of three-centre bridge orbitals replacing the three equivalent orbitals of the triple bond in $C_2H_2$.

## 6.4.5 *Localized $\pi$ orbitals*

It is a feature of localized orbitals that a lone-pair orbital on an atom includes contributions from atomic orbitals on other atoms, and that a bond orbital includes contributions from atomic orbitals on atoms not directly involved in the bond. Two reasons for this residual delocalization are the requirement that the localized orbitals be orthogonal, and that an orbital is not localized in isolation but simultaneously with all the other

138

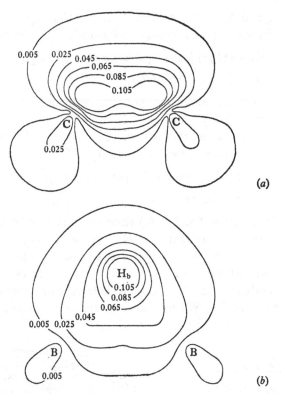

Fig. 6.9. Charge-density contours for localized orbitals in ethylene and diborane: (*a*) a C–C bond orbital in $C_2H_4$, (*b*) a three-centre bridge orbital in $B_2H_6$ (Switkes *et al.* 1969).

orbitals. Both necessarily result in the appearance of small other-atom contributions (see table 6.2). When these contributions are not small however, a meaningful localized-orbital description is not possible. An extreme example of this is provided by the $\pi$ structure of benzene, for which no unique set of localized orbitals can be formed, and the most localized orbitals contain substantial contributions from atomic orbitals on several atoms at a time (Edmiston and Ruedenberg 1966). Thus in a simple LCAO–MO description of the $\pi$ structure, if $\chi_1, \chi_2, ..., \chi_6$ are atomic $\pi$ orbitals on the six carbons of the benzene ring, the delocalized (canonical) molecular orbitals, which are determined wholly by symmetry, are

$$a_{2u} = N_a[\chi_1 + \chi_2 + \chi_3 + \chi_4 + \chi_5 + \chi_6]$$
$$e_{1g} = N_e[\chi_1 \qquad -\chi_3 - \chi_4 \qquad +\chi_6]$$
$$e'_{1g} = N'_e[\chi_1 + 2\chi_2 + \chi_3 - \chi_4 - 2\chi_5 - \chi_6]$$

## The electron distribution

where $N_a$, $N_e$, and $N_e'$ are normalization factors. Corresponding to these there is an infinite number of sets of equally localized equivalent orbitals:

$$\psi_1 = (1/\sqrt{3})\,[a_{2u} + (\sqrt{2})(\cos\alpha\, e_{1g} + \sin\alpha\, e_{1g}')]$$
$$\psi_2 = C_3 \psi_1, \quad \psi_3 = C_3^2 \psi_1$$

where $C_3$ is a rotation through 120° about the principal axis of the molecule, and $\alpha$ is an arbitrary angle.

Of particular interest are the two sets of equivalent orbitals for $\alpha = 0$ and $\alpha = 90°$, of which examples are (from $\psi_1$)

$$\alpha = 0: \quad \pi k = (1/\sqrt{3})\,[a_{2u} + (\sqrt{2})\,e_{1g}]$$
$$\alpha = 90°: \pi l = (1/\sqrt{3})\,[a_{2u} + (\sqrt{2})\,e_{1g}']$$

These are illustrated in fig. 6.10(a) (England and Ruedenberg 1971). We see that the three equivalent orbitals of type $\pi k$ correspond to a single 'Kekulé-type structure', in the sense that each one contains large and equal contributions from two adjacent carbon atoms, and lesser contributions from the other atoms. The second Kekulé-type structure is obtained by rotation through 60°. We note that the two Kekulé-type structures

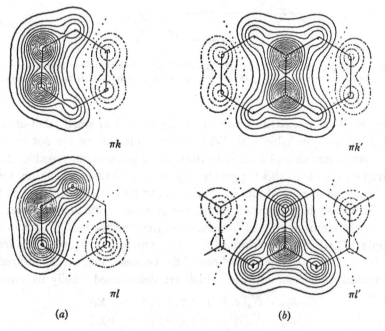

$$\pi k \qquad\qquad\qquad \pi k'$$

$$\pi l \qquad\qquad\qquad \pi l'$$

$$(a) \qquad\qquad\qquad (b)$$

Fig. 6.10. Prototype localized $\pi$ orbitals (England and Ruedenberg 1971).

are merely two equivalent ways of representing the *same* total wave function in terms of approximately two-centre localized $\pi$ orbitals. An alternative representation is provided by the three equivalent orbitals of type $\pi l$, each of which contains a major contribution from one atom only, smaller and equal contributions from the two adjacent atoms, and lesser contributions from the remaining atoms. A second set of these is again obtained by rotation through 60°.

England and Ruedenberg (1971)† have shown that the localized $\pi$ orbitals in condensed hydrocarbons are all basically related to one or more of the four types shown in fig. 6.10. Two of these, $\pi k$ and $\pi k'$, contain major contributions from two adjacent atoms and can be regarded as Kekulé-type whereas the others, $\pi l$ and $\pi l'$, contain major contributions from only one atom. In contrast to the case of benzene, a unique set of localized orbitals is obtained for each condensed hydrocarbon. For example, of the two resonance structures of naphthalene (fig. 6.11), the symmetric structure (*a*) corresponds to the most-localized orbitals. The localized orbitals in naphthalene, phenanthrene and anthracene are illustrated in fig. 6.12. Naphthalene and phenanthrene have Kekulé-type localized structures but anthracene does not. More generally, the condensed hydrocarbons can be divided into two classes: those with a Kekulé-type localized structure, containing localized orbitals of types $\pi k$ and $\pi k'$ only, and those with a non-Kekulé-type localized structure, containing one or more orbitals of types $\pi l$ and $\pi l'$. A partial list of the two classes is given in table 6.4. This division into two classes can be related to the Pauling bond orders of the $\pi$ bonds (England and Ruedenberg 1971), and can be used for an interpretation of those chemical properties of these systems which depend primarily on the $\pi$ structure. It must be remembered however that the localized $\pi$ orbitals have appreciable 'tails' extending over several atoms, showing that there is an appreciable residual delocalization which reduces somewhat the significance of the most-localized structures.

(*a*)                                (*b*)

Fig. 6.11. Resonance structures of naphthalene.

† The Hückel approximation with overlap was used, but the semi-empirical nature of the calculations should not affect the conclusions in any major way.

# The electron distribution

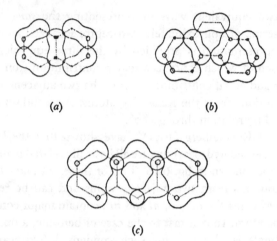

(a)

(b)

(c)

Fig. 6.12. Localized $\pi$ orbitals in (a) naphthalene, (b) phenanthrene, and (c) anthracene (England and Ruedenberg 1971).

TABLE 6.4    *Localized-orbital structures for condensed hydrocarbons (England and Ruedenberg 1971)*

| Kekulé-type | Non-Kekulé-type |
| --- | --- |
| Naphthalene | Azulene |
| Triphenylene | Anthracene |
| Phenanthrene | Naphthacene |
| 1,2,3,4-dibenzanthracene | Pentacene |
| Benzphenanthrene | 1,2,5,6-dibenzanthracene |
| Benzanthracene | 1,2,7,8-dibenzanthracene |
| Chrysene | Pyrene |
| Picene | Benzpyrene |
| Pentaphene | Anthanthrene |
| Perylene | |
| Benzoperylene | |
| Coronene | |

## 6.4.6  *Hybridization*

The transformation of (canonical) molecular orbitals to localized orbitals provides a link between the formal molecular-orbital theory and the chemist's picture of the electronic structure in terms of bonds and lone pairs, whose interpretation has often invoked the concept of orbital hybridization.

A hybrid orbital on an atom A is an atomic orbital which is a linear

142

combination of one-centre functions on A with different symmetry properties: $h_A = c_s S_A + c_p P_A + c_d D_A + \ldots$

where $S_A$, $P_A$, $D_A$, ... are normalized atomic orbitals of types $s$, $p$, $d$, ..., each of which may be a linear combination of several more elementary functions of the same symmetry type. The extent of hybridization is given by the ratios of the squares of the coefficients: $s^i p^j d^k \ldots$ where $i:j:k \ldots = c_s^2 : c_p^2 : c_d^2 \ldots$. Such hybrid orbitals may be extracted from a computed LCAO–MO wave function by an examination of the localized orbitals. Thus, a localized orbital describing a two-centre bond between atoms A and B has the form

$$\psi_{AB} = c_A h_A + c_B h_B + \text{other-atom contributions}$$

Two properties of the hybrid orbitals are of interest: the extent of hybridization and the direction of maximum density. In the localized orbitals of $H_2O$ given in table 6.2, the hybridization is $sp^{1.5}$ for the lone-pair orbitals and $sp^{6.6}$ for the bond orbitals. The angle between the bond hybrids is about 90°, compared with the molecular bond angle of 104.5°, and the angle between the lone-pair hybrids is 124°. These results are consistent with a simple picture of the electronic structure in which the bonding is interpreted as arising from the overlap of essentially pure $p$-type orbitals on the oxygen and the $1s$ orbitals on the hydrogens. The difference of 15° between the observed value of the bond angle and the value of 90° is then accounted for by the repulsion of the fractional positive charges on the hydrogens.

Bond orbitals may in general be expected to be bent, with the maximum density displaced from the bond axis. In many molecules however this bending is small except for multiple bonds described by equivalent orbitals, or when the structure is 'strained' as in cyclopropane. The hybridization of the localized C–H bond orbital in different molecular environments is summarized in table 6.5. The orbitals have been obtained by an Edmiston–Ruedenberg localization of SCF molecular orbitals constructed from minimum bases of Slater functions (Newton, Switkes and Lipscomb 1970b; Newton and Switkes 1971). The deviations of the hybrids from the bond directions are small, and of little significance. The hybridization is consistent with the simple picture of the bonding in terms of $sp^3$, $sp^2$ and $sp$ hybrids in these molecules, but there are some obvious differences of detail. For example, the hybridization in the saturated hydrocarbons is about $sp^{2.7}$ instead of $sp^3$ as required for classical tetrahedral hybrids. One reason for this difference is that the computed

TABLE 6.5 *Hybridization in the C–H bond orbital. In the case of mole-cules containing multiple bonds, the values refer to the $\sigma$–$\pi$ description of the bonding (Newton et al. 1970b; Newton and Switkes 1971)*

| Molecule | Hybrid ($sp^x$) $x$ | Angle with bond direction |
|---|---|---|
| $CH_4$ | 2.65 | 0.0° |
| $C_2H_6$ (staggered) | 2.72 | 1.0° |
| $C_2H_6$ (eclipsed) | 2.75 | 1.5° |
| $C_2H_4$ | 1.85 | 1.2° |
| $CH_2{=}C'H{-}C'H{=}CH_2$ (*trans*) | 1.87 (C'H) | < 3.0° |
| | 1.79 (CH) | < 3.0° |
| Benzene | 1.87 | 0.0° |
| $C_2H_2$ | 1.01 | 0.0° |
| HCN | 0.89 | 0.0° |
| Cyclopropane | 2.04 | 1.9° |

hybrids, unlike the classical hybrids, are not required to be orthogonal, so that there is no constraint on the types of hybridization possible for a given set of inter-hybrid angles.

In the case of molecules containing multiple bonds, the values in table 6.5 refer to the $\sigma$–$\pi$ description of the bonding. The hybridization in the C–C $\sigma$ bond is $sp^{1.74}$ for $C_2H_4$ and $sp^{1.01}$ for $C_2H_2$, and these values are fairly typical of the hybridization in double and triple bonds. In the fully localized description, the C–H bond orbitals are almost unchanged but the mixing of the $\sigma$ and $\pi$ orbitals of the multiple bonds leads to equiva-lent C–C bond hybrids with high $p$-contents: $sp^{4.8}$ in $C_2H_4$ and $sp^{5.8}$ in $C_2H_2$. In $C_2H_4$ the hybrids make an angle of 56° with the C–C axis. In $C_2H_2$ the angle is 68° and this value is characteristic of the triple bond: 66° in $N_2$, 66° for the carbon hybrid in CO, and 68° for the oxygen hybrid in CO. In cyclopropane (Newton *et al.* 1970b) the C–C bond hybrids again have a high $p$-content ($sp^{3.8}$), and they deviate from the C–C direction by 28°. The bulk of the corresponding orbital density lies outside the ring, as shown in fig. 6.13. A consideration of the total charge-density distribution (Kochanski and Lehn 1969; Stevens *et al.* 1971) suggests however that both the bending of the C–C bonds and the displacement of charge out of the ring are not as great as implied in fig. 6.13. The 'bent-bond plus central-hole' picture of the molecule is confirmed, but the direction of maximum density of the C–C bond devi-ates by less than 10° from the C–C direction, and the minimum density in the hole is only about 20 per cent below this maximum. These results

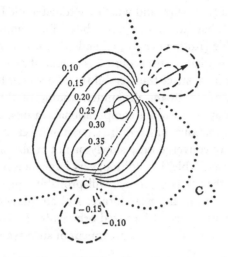

Fig. 6.13. Localized orbital for the C–C bond in cyclopropane. ↔ indicates the hybrid direction, and —·—·— the internuclear direction (Newton, Switkes and Lipscomb 1970*b*).

demonstrate that a consideration of the individual orbitals only can give an incomplete and sometimes misleading description of the overall electron distribution.

The set of localized orbitals obtained by the Edmiston–Ruedenberg procedure is only one of many different sets which can be formed by unitary transformations of the canonical molecular orbitals. Each set provides a different interpretation of the bonding. Orbitals obtained by different procedures, for example, can exhibit quite different degrees of hybridization and inter-hybrid angles. Thus, it is possible to form localized bond orbitals for the water molecule in which the oxygen hybrids are approximately $sp^3$ and point along the bond axes (McWeeny and Del Re 1968). The resulting picture of the bonding is as valid as, and simpler than, that obtained in terms of the Edmiston–Ruedenberg orbitals. This arbitrariness of the orbitals, and of the hybridization, partially explains the success of the various qualitative interpretations of molecular bonding based on chemical intuition and experience, since it is often possible to find a transformation of the orbitals which is consistent with any particular interpretation.

Much of the interest of this type of analysis lies in the possibility of finding localized orbitals and hybrid orbitals which are characteristic of

## The electron distribution

certain bonds and lone pairs, and which are transferable from one molecule to another. That the C–H bond orbital, for example, is approximately transferable (Rothenberg 1969, and table 6.5) is not a very surprising result, and what is still required is a quantitative understanding of the deviations from strict transferability. The search for transferable orbitals is important both for the understanding of the properties of related molecules and for various computational reasons. Thus, the solution of the SCF equations may require a large number of iterations, and the procedure may converge only slowly or not at all unless very good initial orbitals are available. In addition, very large molecules, containing hundreds rather than tens of bonds, will probably not be amenable to accurate computation in the near future, and an understanding of the transferability or otherwise of localized orbitals for individual bonds or groups of bonds is essential for the treatment of such systems.

### 6.4.7 Localized correlated pair functions

We saw in §4.3.1 that the total correlation energy of a molecule can often be approximated as a sum of pair energies, (4.18),

$$E_C \simeq \sum_{k<l} e(k,l)$$

where $e(k,l)$ is the correlation energy associated with a pair of electrons in the spin-orbitals $\phi_k$ and $\phi_l$. The corresponding approximate correlated wave function has the form (4.16).

$$\Psi = \Psi_{RHF} + \sum_{k<l} \Phi_{kl}$$

in which $\Psi_{RHF}$ is an orbital approximation,

$$\Psi_{RHF} = (N!)^{-\frac{1}{2}} \det[\phi_1(1)\,\phi_2(2) \ldots \phi_N(N)]$$

and $\Phi_{kl}$ is a pair-correlated wave function which is obtained from $\Psi_{RHF}$ by replacing the pair of spin-orbitals $\phi_k$ and $\phi_l$ by a correlated pair function $\phi_{kl}$; for example,

$$\Phi_{1,2} = (N!)^{-\frac{1}{2}} \det[\phi_{1,2}(1,2)\,\phi_3(3) \ldots \phi_N(N)]$$

It is found in general that when $\Psi_{RHF}$ is constructed from a set of delocalized molecular orbitals, the corresponding pair functions are also delocalized. On the other hand, a transformation to localized orbitals

146

produces new pair functions which are localized in the same regions of the space as the orbitals from which they are derived. A transformation of the orbitals also leads to a new partitioning of the total correlation energy $E_C$. The valence-shell pair-correlation energies for the delocalized (canonical) molecular orbitals and the localized orbitals in the ground state of $CH_4$ are compared in table 6.6 (Meyer 1973). The localization of the orbitals produces intrabond pair functions which are localized about the bond axes, and the close proximity of the two electrons in the same bond orbital gives rise to a greatly enhanced intrapair correlation energy of $0.027H_\infty$ per bond, which is of comparable magnitude to the inner-shell pair energy of $0.036H_\infty$. The total interpair correlation energy is correspondingly decreased.

TABLE 6.6 *Comparison of valence-shell pair-correlation energies for canonical and localized orbitals in the ground state of* $CH_4$. *The figures in brackets are the numbers of spin-orbital pairs corresponding to the same orbital pair. The pair energies are totals for each type of orbital pair* (*Meyer 1973*)

| Orbital pair | | Pair energy/$H_\infty$ |
|---|---|---|
| | CANONICAL ORBITALS | |
| $(2a_1)^2$ | | 0.009 (1) |
| $(1t_2)^2$ | | 0.042 (3) |
| $(2a_1, 1t_2)$ | | 0.053 (12) |
| $(1t_2, 1t_2')$ | | 0.076 (12) |
| | Total intrapair energy | 0.051 |
| | Total interpair energy | 0.129 |
| | Intra/inter = 0.393 | |
| | LOCALIZED ORBITALS | |
| $(bCH)^2$ | | 0.109 (4) |
| $(bCH, bCH')$ | | 0.070 (6) |
| | Intra/inter = 1.55 | |

The importance of localization in the study of electron correlation lies, as for the orbitals themselves, in the possible transferability of localized pair functions and pair energies from one molecule to another. A knowledge of the pair energies characteristic of localized orbitals and pairs of orbitals can provide an estimate of the total correlation energy of a molecule, and of the change in correlation energy which accompanies a change of state.

# The electron distribution

## 6.5 POPULATION ANALYSIS

### 6.5.1 Atomic populations, overlap populations and effective atomic charges

In §6.3.1 we considered the partitioning of the total charge distribution into contributions associated with the different occupied molecular orbitals. These orbital contributions normally span several atoms in a molecule, unless the orbitals are highly localized, and it is sometimes more convenient for interpretative purposes to consider a different type of partitioning in which the separate contributions are associated with the individual atoms in the molecule. Such a partitioning can then be used to define effective 'atomic charges', and these provide a simple picture of the changes in the electron distribution that accompany, for example, an excitation or ionization, the formation of a bond, or a chemical reaction.

The simplest and most widely used method of calculating atomic charges is the electron population analysis due to Mulliken (1955), which involves a partitioning of the total molecular charge distribution into atomic-orbital contributions, and can be used for any molecular wave function which has been expressed in terms of atomic orbitals centred on the nuclei.

Let $A(r)$ be an atomic orbital centred on a nucleus A of a diatomic molecule AB, and $B(r)$ an atomic orbital on nucleus B. Let $\psi_i$ be a molecular orbital which is a combination of the two atomic orbitals:

$$\psi_i(r) = a_i A(r) + b_i B(r)$$

The corresponding molecular-orbital density function is

$$P^{(i)}(r) = |\psi_i(r)|^2 = P_A^{(i)}(r) + P_B^{(i)}(r) + P_{AB}^{(i)}(r) \qquad (6.28)$$

where $\quad P_A^{(i)}(r) = a_i^* a_i A(r)^* A(r), \quad P_B^{(i)}(r) = b_i^* b_i B(r)^* B(r)$

$$P_{AB}^{(i)}(r) = a_i^* b_i A(r)^* B(r) + b_i^* a_i B(r)^* A(r)$$

The quantity $P_A^{(i)}$, involving the atomic orbital $A$ only, is an *atomic density*, and is interpreted as that part of the molecular–orbital density which is to be associated with atom A. The quantity $P_{AB}^{(i)}$ is an *overlap density* (or bond density), and is interpreted as that part of the orbital density which is to be associated with the bond linking the atoms A and B. More generally, let $A_p(r)$ ($p = 1, 2, ..., N_A$) be a set of atomic orbitals centred on nucleus A and let $B_s(r)$ ($s = 1, 2, ..., N_B$) be a set of atomic orbitals on nucleus B. The set of $(N_A + N_B)$ atomic orbitals can be used

148

to construct $(N_A + N_B)$ orthonormal molecular orbitals,

$$\psi_i(\mathbf{r}) = \sum_{p=1}^{N_A} a_{ip} A_p(\mathbf{r}) + \sum_{s=1}^{N_B} b_{is} B_s(\mathbf{r}) \qquad (6.29)$$

whose orbital densities have the form (6.28) with, for example,

$$P_A^{(i)}(\mathbf{r}) = \sum_{p,\,q=1}^{N_A} a_{ip}^* a_{iq} A_p(\mathbf{r})^* A_q(\mathbf{r}) \qquad (6.30)$$

$$P_{AB}^{(i)}(\mathbf{r}) = \sum_{p=1}^{N_A} \sum_{s=1}^{N_B} [a_{ip}^* b_{is} A_p(\mathbf{r})^* B_s(\mathbf{r}) + b_{is}^* a_{ip} B_s(\mathbf{r})^* A_p(\mathbf{r})] \qquad (6.31)$$

For a polyatomic molecule,

$$P^{(i)}(\mathbf{r}) = \sum_\alpha P_\alpha^{(i)}(\mathbf{r}) + \sum_{\alpha>\beta} \sum P_{\alpha\beta}^{(i)}(\mathbf{r}) \qquad (6.32)$$

in which the summations are over the atoms and pairs of atoms in the molecule.

In the orbital approximation (or more generally, when the $\psi_i$ are natural orbitals) the total molecular density function is the sum of these molecular-orbital contributions:

$$P(\mathbf{r}) = \sum_i n_i P^{(i)}(\mathbf{r}) \qquad (6.33)$$

where $n_i$ is the occupation number of orbital $\psi_i$ $(n_i \leqslant 2)$, and this may now also be partitioned into contributions from the atoms and pairs of atoms:

$$P(\mathbf{r}) = \sum_\alpha P_\alpha(\mathbf{r}) + \sum_{\alpha>\beta} \sum P_{\alpha\beta}(\mathbf{r}) \qquad (6.34)$$

where $\qquad P_\alpha(\mathbf{r}) = \sum_i n_i P_\alpha^{(i)}(\mathbf{r}), \quad P_{\alpha\beta}(\mathbf{r}) = \sum_i n_i P_{\alpha\beta}^{(i)}(\mathbf{r}) \qquad (6.35)$

are total atomic and total overlap density functions.

The partitionings (6.32) and (6.34) of the orbital and total density functions form the basis of the Mulliken population analysis. Thus, the $n_i$ electrons in the molecular orbital $\psi_i$ can be partitioned amongst the atoms and bonds in accordance with (6.32) by means of the identity

$$n_i \equiv n_i \int P^{(i)}(\mathbf{r}) \, dv = \sum_\alpha n_i(\alpha) + \sum_{\alpha>\beta} \sum n_i(\alpha\beta) \qquad (6.36)$$

where $\qquad n_i(\alpha) = n_i \int P_\alpha^{(i)}(\mathbf{r}) \, dv, \quad n_i(\alpha\beta) = n_i \int P_{\alpha\beta}^{(i)}(\mathbf{r}) \, dv \qquad (6.37)$

The quantity $n_i(\alpha)$ is the number of electrons in the molecular orbital $\psi_i$ which is associated with the atom $\alpha$ in the molecule, and is called an

## The electron distribution

*orbital net atomic population*. The quantity $n_i(\alpha\beta)$ is the number of electrons which is associated with the pair of atoms $\alpha$ and $\beta$, and is an *orbital overlap population*. For a diatomic molecule with orbitals (6.29), the orbital populations are obtained by integration of the densities (6.30) and (6.31):

$$n_i(\text{A}) = n_i \sum_{p,\,q=1}^{N_A} a_{ip}^* a_{iq} S_{pq}$$

$$n_i(\text{AB}) = n_i \sum_{p=1}^{N_A} \sum_{s=1}^{N_B} [a_{ip}^* b_{is} S_{ps} + b_{is}^* a_{ip} S_{sp}]$$

where

$$S_{pq} = \int A_p^* A_q \, \mathrm{d}v, \quad S_{ps} = \int A_p^* B_s \, \mathrm{d}v$$

are overlap integrals. The orbital populations may be combined to give total net atomic and total overlap populations:

$$n(\alpha) = \sum_i n_i(\alpha) = \int P_\alpha(\mathbf{r}) \, \mathrm{d}v \qquad (6.38)$$

$$n(\alpha\beta) = \sum_i n_i(\alpha\beta) = \int P_{\alpha\beta}(\mathbf{r}) \, \mathrm{d}v \qquad (6.39)$$

where $P_\alpha$ and $P_{\alpha\beta}$ are the total atomic and overlap densities (6.35). In addition, a number of sub-totals may also be defined by restricting the summations in (6.38) and (6.39) to certain classes of orbitals. Obvious examples are the $\sigma$ and $\pi$ populations in molecules containing both types of orbitals.

The net atomic and overlap populations may be further combined to give a partitioning of the total electron population of the molecule amongst the atoms only. This requires a separation of each overlap population into two parts associated with the two relevant atoms. The Mulliken prescription is to assign exactly half of the overlap population to each atom, and the resulting quantities,

$$N_i(\alpha) = n_i(\alpha) + \tfrac{1}{2} \sum_{\beta \neq \alpha} n_i(\alpha\beta) \qquad (6.40)$$

$$N(\alpha) = \sum_i N_i(\alpha) \qquad (6.41)$$

are orbital and total *gross populations* on atom $\alpha$, and

$$N = \sum_\alpha N(\alpha)$$

is the total number of electrons in the molecule. The quantity $-eN(\alpha)$

can then be interpreted as the electronic charge on atom $\alpha$ and, if the nuclear charge is $Z_\alpha e$, the total charge (sometimes called the net charge) on the atom is

$$q(\alpha) = [Z_\alpha - N(\alpha)]\, e \qquad (6.42)$$

Corresponding gross populations and charges may also be defined, for example, for the $\sigma$ and $\pi$ distributions.

The Mulliken populations and charges are readily calculated for a wave function of the LCAO–MO type, and for any more general wave function which has been expressed in terms of a basis of atomic orbitals centred on the nuclei. It is clear however that the results of such a population analysis must depend to some extent on the numbers and types of atomic orbitals, and we shall consider the validity and limitations of the method in §6.5.3. We consider first some representative results.

**6.5.2** *Some examples of population analysis*
Most commonly used for interpretative purposes are the gross populations $N(\alpha)$, or the corresponding charges $q(\alpha)$, and the overlap populations $n(\alpha\beta)$. In fig. 6.14 are shown the total charges and the separate contributions of the $\sigma$ and $\pi$ distributions obtained from an LCAO–MO calculation for the ground state of pyrrole (Clementi, Clementi and Davis 1967). The $\sigma$ distribution is that expected from a consideration of the relative electronegativities of the atoms, whilst the $\pi$ distribution is similar to that obtained from resonance theory (Pauling 1960) or from $\pi$-electron molecular-orbital theory (Coulson 1961). The nitrogen is seen to be a $\pi$-donor and a $\sigma$-acceptor, with a resultant negative charge. This two-way flow of charge is frequently observed in molecules containing both $\sigma$ and $\pi$ bonds, and it represents a competition between inductive and conjugative or mesomeric effects.

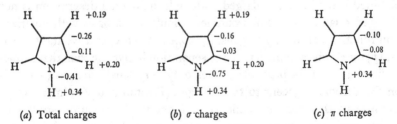

(*a*) Total charges     (*b*) $\sigma$ charges     (*c*) $\pi$ charges

Fig. 6.14. Total, $\sigma$ and $\pi$ atomic charges (in units of $e$) in the ground state of pyrrole (Clementi, Clementi and Davis 1967).

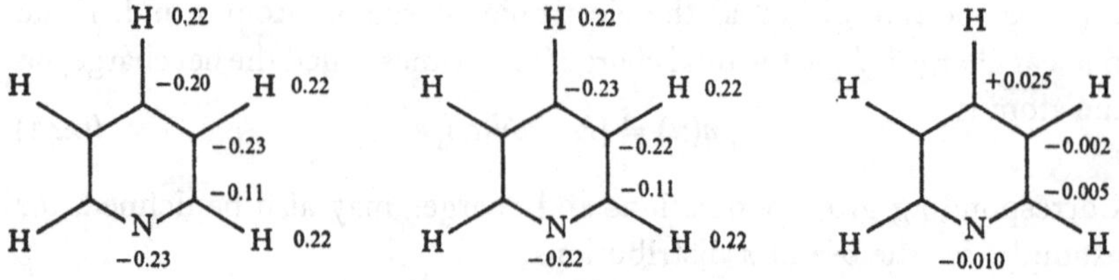

# The electron distribution

Fig. 6.15. Total, $\sigma$ and $\pi$ atomic charges (in units of $e$) in the ground state of pyridine (Clementi 1967a).

A somewhat different trend is found in pyridine. The charges in fig. 6.15 (Clementi 1967a) show that the overall charge distribution is dominated by the $\sigma$ structure, with very little sign of the movement of $\pi$ charge onto the nitrogen which is expected from $\pi$-electron theory. This type of result has helped to cast serious doubt on the ability of $\pi$-electron theory to give even qualitatively meaningful results. Further evidence of this is obtained from the reorganization of the electron distribution that accompanies ionization in $\pi$-electron systems. Thus, the removal of an electron from the highest occupied $\pi$ orbital, $1a_2$, in pyridine is accompanied by a greater redistribution of $\sigma$ charge than of $\pi$ charge (Clementi 1967b). The $\pi$ charge is removed mainly from the *ortho*- and *meta*-carbons (about a quarter of an electron from each of these), but this loss is partially neutralized by a transfer of $\sigma$ charge from the hydrogens into the ring, the *ortho*- and *meta*-carbons each receiving about 0.15 electrons. Similarly, the removal of an inner-shell electron is accompanied by a large migration of both $\sigma$ and $\pi$ charge towards the positive hole of the ion (Clark and Scanlan 1974).

The charges shown in figs. 6.14 and 6.15 have been obtained from SCF calculations in terms of the minimal contracted Gaussian basis of $(7s, 3p/2s, 1p)$ on C and N, and $(3s/1s)$ on H. Somewhat different charges are obtained if the basis is enlarged, although the overall description is not changed qualitatively. For example, a double-zeta basis with polarization functions on the hydrogens, $(9s,5p/4s, 2p)$ on C and N, and $(4s, 1p/2s, 1p)$ on H, gives a smaller charge separation in pyrrole than that shown in fig. 6.14 and, in particular, gives only very small charges $(+0.01e)$ on the carbons adjacent to the nitrogen (Preston and Kaufman 1973). Because of this basis-dependence, comparative population studies of related molecules, or of different geometries of the same molecule, are usually more instructive and significant than isolated studies, since the predicted trends can then be expected to be rather insensitive to the way the wave functions have been calculated. In table 6.7 are summarized the atomic charges and overlap populations for borane and the fluorinated boranes (Schwartz and Allen 1970). The total charges on H and F are essentially constant throughout the series, but the charge on B changes linearly from $-0.57e$ in $BH_3$ to $+1.39e$ in $BF_3$. We observe again a two-way transfer of charge, fluorine being the $\sigma$-acceptor and $\pi$-donor. The distribution of charge is also reflected by the overlap populations. The B–H bond is insensitive to fluorination of the molecule, with an overlap population of about 0.8 which is fairly typical of non-polar covalent bonds. The smaller values for the B–F bond reflect the greater ionic character of this bond, and the greater difference of the electronegativities of the bonded atoms,.

TABLE 6.7   *Atomic charges (in units of e) and overlap populations in borane and the fluorinated boranes (Schwartz and Allen 1970)*

| | ATOMIC CHARGES | | | | | | |
| | B | | | H | F | | |
| | $\sigma$ | $\pi$ | Total | Total ($\sigma$) | $\sigma$ | $\pi$ | Total |
|---|---|---|---|---|---|---|---|
| $BH_3$ | −0.569 | 0 | −0.569 | +0.190 | | | |
| $BH_2F$ | +0.280 | −0.198 | +0.082 | +0.179 | −0.637 | +0.198 | −0.439 |
| $BHF_2$ | +1.052 | −0.326 | +0.726 | +0.187 | −0.620 | +0.163 | −0.457 |
| $BF_3$ | +1.803 | −0.415 | +1.388 | | −0.601 | +0.138 | −0.463 |

| | OVERLAP POPULATIONS | | | |
| | B–H | B–F | | |
| | Total ($\sigma$) | $\sigma$ | $\pi$ | Total |
|---|---|---|---|---|
| $BH_3$ | 0.794 | | | |
| $BH_2F$ | 0.806 | 0.223 | 0.204 | 0.426 |
| $BHF_2$ | 0.825 | 0.294 | 0.171 | 0.465 |
| $BF_3$ | | 0.378 | 0.147 | 0.525 |

A small value of an overlap population indicates either that the bond is essentially ionic or that no conventional 'bond' exists between the atoms. Thus the overlap population of the bond of LiF is 0.26 (Williams and Streitwiesser 1974). Another example of ionic bonding is provided by the octahedral complex $NiF_6^{4-}$ (Moskowitz *et al.* 1970) in which the Ni–F overlap population is $-0.01$ and the atomic charges are $1.82e$ on Ni and $-0.97e$ on F. The system consists essentially of the ions $Ni^{2+}$ and $F^-$ bound by electrostatic forces with some ligand-to-metal back-donation. The small overlap population shows that covalency effects are very small, and the negative value of the overlap population, although possibly not significant, suggests that the covalency effects are antibonding. The atomic populations are further analysed in table 6.8 in terms of the separate contributions of the atomic orbitals. The atomic-orbital populations are seen to be almost identical to those in the free ions, with little promotion into the Ni $4s$ and $4p$ orbitals.

TABLE 6.8   *Gross atomic populations in* $NiF_6^{4-}$ *(Moskowitz et al. 1970)*

|  | Ni | F |
|---|---|---|
| $1s$ | 2.000 | 1.999 |
| $2s$ | 1.999 | 1.985 |
| $2p$ | 6.000 | 5.986 |
| $3s$ | 1.998 | |
| $3p$ | 5.993 | |
| $3d$ | 8.058 | |
| $4s$ | 0.071 | |
| $4p$ | 0.062 | |
| Total | 26.181 | 9.970 |

### 6.5.3   *The basis-dependence of population analysis*

We turn now to a consideration of the validity of the Mulliken population analysis. It has already been suggested that whereas the molecular charge-density function, from which the populations are derived, is in principle independent of the way it has been calculated, its partitioning into atomic and overlap densities, as given by (6.34), is highly sensitive to the way the wave function has been constructed; that is, to the num-bers and types of atomic orbitals used as basis. Thus, it is in principle possible to calculate a molecular wave function to any desired accuracy in terms of a sufficiently large basis of atomic orbitals all of which are centred on the same arbitrary point. If this centre lies on a nucleus then all the electronic charge is presumably associated with one atom, and a population analysis would give net and gross populations on this atom equal to the total number of electrons in the molecule, and zero values on all other atoms. This is an extreme example of the use of an 'un-balanced basis' (Mulliken 1962). A balanced basis is one for which the numbers and types of atomic orbitals allow all the atoms to make equally flexible and balanced contributions to the delocalization, polarization and contraction effects that accompany the formation of a molecule (see chapter 7). Such a basis might consist of a double-zeta set of primary atomic orbitals plus a set of polarization functions on *each* atom, including hydrogen, but there are no simple general rules for the construction of balanced bases and, as we shall see, even the use of apparently well-balanced bases does not necessarily guarantee consistent results. The (Mulliken) population analysis may also be criticized for the equal division (6.40) of the overlap populations between pairs of atoms. Such a division is clearly unrealistic, unless the atoms are equivalent, particularly when the centroid of the overlap distribution lies much closer to one nucleus than to the other. This may be one reason why, for example, the charge on hydrogen in molecules such as $BH_3$ (table 6.7) is nearly always found to be positive, despite the small size of the hydrogen and, in the case of $BH_3$, the similar electronegativities of the atoms.

An example of the basis-dependence of population analysis is provided by a comparison (Kern and Karplus 1964) of the electron distribution predicted for hydrogen fluoride by two equally accurate but independent SCF calculations. The basis sets used for the construction of the two wave functions are large and apparently well-balanced sets of 18 Slater functions (Nesbet 1962) and 16 Slater functions (Clementi 1962). The calculated total energies are $-100.0571H_\infty$ and $-100.0575H_\infty$ respectively, compared with the RHF limiting value of $-100.07H_\infty$ (Cade and Huo 1967a). The corresponding charge-density distributions are also very similar, but the population analyses give significantly different atomic charges: $-0.23e$ on F for the Nesbet wave function, and $-0.48e$ for the Clementi wave function. Such inconsistent results can be avoided in general only by computing the atomic charges directly from the charge-density function, by a suitable partitioning of the molecular space into regions associated with the individual atoms, and integration of the charge density inside each region. Such a partitioning is of course still necessary arbitrary, but it has the advantage that the resulting charges are no longer dependent on the mathematical form of the wave function.

Politzer and Harris (1970) have proposed a partitioning of the molecular space in which the atomic regions are defined such that in the limit of no interactions between the atoms, the electronic charge associated with each atom in the molecule is the same as that of the free atoms. Thus, for a diatomic molecule like HF, the boundary between the regions is determined by first placing the non-interacting atoms at the appropriate positions for the molecule and then dividing the molecular space into two

## The electron distribution

regions by means of a plane perpendicular to the molecular axis, such that each region has the appropriate number of electrons: one for hydrogen and nine for fluorine. The interaction of the atoms then results in a redistribution of charge amongst the regions so defined. Application of this method to the Nesbet and Clementi wave functions for HF (Politzer and Mulliken 1971) gives a charge of $-0.27e$ on F for the Nesbet wave function and $-0.26e$ for the Clementi wave function, compared to $-0.23e$ and $-0.48e$, respectively, obtained from population analysis.

TABLE 6.9 *Atomic charges (in units of e) in methane and the fluorinated methanes calculated from two sets of wave functions*

| Molecule | (i) Minimal STO–3G | | | (ii) Double-zeta basis of contracted Gaussians | | |
|---|---|---|---|---|---|---|
| | H | C | F | H | C | F |
| $CH_4$ | +0.018 | −0.073 | | +0.185 | −0.739 | |
| $CH_3F$ | −0.004 | +0.169 | −0.157 | +0.174 | −0.140 | −0.383 |
| $CH_2F_2$ | −0.023 | +0.383 | −0.169 | +0.175 | +0.327 | −0.338 |
| $CHF_3$ | −0.018 | +0.532 | −0.171 | +0.184 | +0.665 | −0.283 |
| $CF_4$ | | +0.674 | −0.169 | | +0.936 | −0.234 |

References: for (i) Hehre and Pople 1970; for (ii) Brundle, Robin and Basch 1970.

The foregoing discussion suggests that the Mulliken population analysis can sometimes give a quite misleading picture of the distribution of charge in a molecule. On the other hand, there is some evidence that the relative values of the charges in a molecule or in a series of related molecules do often reproduce the essential features and trends of the electron distribution, and that they can then be correlated with the relative values of other properties. But here again, the use of different types of wave functions can lead to inconsistent results. In table 6.9 are shown the (Mulliken) atomic charges calculated from two different sets of wave functions for methane and the fluorinated methanes. The two sets of charges are clearly very different, the charge separation for the (energetically) more accurate double-zeta basis (ii) being much larger in all cases than that for the minimal basis (i). Although little significance can therefore be attached to the magnitudes, or even to the signs, of the charges in any one molecule, there is some agreement on the trends. The trend predicted for the carbon charges is the same for both bases, and there is a good linear correlation between the two sets of charges. On the other hand, whereas the fluorine charges (ii) become less negative with increased number of fluorines, the reverse, and possibly less reasonable, trend is shown by the set (i). Because of these differences, no significance can be attached to the small changes predicted for the hydrogen charges (Naleway and Schwartz 1973).

TABLE 6.10 *Energies, atomic charges and d-orbital populations in* $SiH_4$, $PH_3$, $H_2S$ *and* $HCl$ *for three sets of basis functions*

| Molecule | Property | (i) Minimal STO | (ii) Minimal STO$+d$ on heavy atom | (iii) Extended GTO including $d$ on heavy atom and $p$ on H |
|---|---|---|---|---|
| $SiH_4$ | Energy/$H_\infty$ | $-290.4250$ | $-290.5197$ | $-291.2355$ |
|  | $q$ (Si)/$e$ | $-0.220$ | $-0.508$ | $+0.803$ |
|  | $q(H)/e$ | $+0.055$ | $+0.127$ | $-0.201$ |
|  | $d$ population |  | $0.447$ | $0.108$ |
| $PH_3$ | $q(P)/e$ |  |  | $+0.244$ |
|  | $q(H)/e$ |  |  | $-0.081$ |
|  | $d$ population |  |  | $0.082$ |
| $H_2S$ | Energy/$H_\infty$ | $-397.7881$ | $-397.8415$ | $-398.6862$ |
|  | $q(S)/e$ | $-0.186$ | $-0.352$ | $-0.098$ |
|  | $q(H)/e$ | $+0.093$ | $+0.176$ | $+0.049$ |
|  | $d$ population |  | $0.147$ | $0.063$ |
| $HCl$ | $q(Cl)/e$ |  |  | $-0.177$ |
|  | $q(H)/e$ |  |  | $+0.177$ |
|  | $d$ population |  |  | $0.030$ |

References: for (i) and (ii) Boer and Lipscomb 1969; for (iii) Rothenberg, Young and Schaefer 1970.

The charges shown in table 6.9 are for wave functions which have been calculated in terms of basis sets containing primary atomic orbitals only: $s$ and $p$ on C and F, $s$ on H. The effects of adding polarization functions are summarized in table 6.10 for some second-row hydrides. All three sets show the expected uniform decrease of electron population on the hydrogens as the electronegativity of the heavy atom increases. The addition of a set of $d$-type polarization functions to the minimal STO basis on Si and S results in a marked increase of electron population on the heavy atom at the expense of the hydrogens, with a substantial population of the $d$ orbitals, particularly in $SiH_4$. The further extension of the basis to include $p$-type polarization functions on the hydrogens, as well as more primary $s$ and $p$ functions, results in a decrease in the importance of the $d$ orbitals, and increased population on the hydrogens. The negative charges on the hydrogens in $SiH_4$ for example are now more consistent

# The electron distribution

with the relative electronegativities and sizes of the atoms. One reason for the smaller population of the $d$ orbitals, 0.11 as compared with 0.45 in $SiH_4$, is that these orbitals are no longer used to compensate for the deficiencies of the minimal $s$ and $p$ basis on the heavy atom, and this is reflected by a correspondingly smaller contribution of the $d$ orbitals to the total energy: $-0.03H_\infty$ for basis (iii) as compared with $-0.09H_\infty$ for basis (ii) (Rothenberg, Young and Schaefer 1970). A second reason is the presence of the $p$-type polarization functions which allow the hydrogens to make a more balanced contribution to the bonding.

Information as to the nature of the bonding in a molecule is provided by the overlap populations, but these can again be highly sensitive to the nature of the basis, as shown by the values for LiF in table 6.11 (Williams and Streitwieser 1974). The values of the dipole moment and force constant show that the two smallest basis sets, 1 and 2, do not provide accurate descriptions of the electron distribution. The presence of only a minimal set of primary atomic orbitals in basis 1 exaggerates the importance of the lithium $p$ orbitals, and therefore inhibits the transfer of charge from the lithium onto the fluorine. This results in small atomic charges and dipole moment, and large overlap population and force constant which are more characteristic of covalent bonding than of ionic bonding. On the other hand, the absence of lithium $p$ orbitals from basis 2 enhances the lithium-to-fluorine transfer of charge, and results in an almost pure ionic bond. The addition to basis 2 of a lithium $p\sigma$ orbital has relatively

TABLE 6.11  *Total energy* $E$, *dipole moment* $\mu$, *atomic charge* $q(Li)$, *overlap population* $n(LiF)$ *and force constant* $k$ *for the ground state of* LiF (*Williams and Streitwieser 1974*)

| Basis | $E/H_\infty$ | $\mu/D$ | $q(Li)/e$ | $n(LiF)$ | $k/N\,cm^{-1}$ |
|---|---|---|---|---|---|
| 1 Minimal $+p$(Li) | $-106.1299$ | 3.91 | 0.300 | 0.484 | 5.63 |
| 2 Double-zeta | $-106.8907$ | 7.26 | 0.977 | $-0.035$ | 3.71 |
| 3 $2+p\sigma$(Li) | $-106.8963$ | 7.11 | 0.914 | 0.023 | 3.30 |
| 4 $3+p\pi$(Li) | $-106.9225$ | 6.36 | 0.754 | 0.268 | 3.30 |
| 5 $4+d$(F) | $-106.9292$ | 6.10 | 0.761 | 0.260 | 3.03 |
| Hartree–Fock | $-106.9916$ | 6.30 | | | 2.57 |
| Experiment | $-107.504$ | 6.28 | | | 2.82 |

Basis 1 is a minimal STO-4G basis plus a set of $2p$ orbitals on Li. Basis 2 is the contracted Gaussian basis Li($8s/4s$) and F($8s$, $4p/4s$, $2p$). Bases 3, 4 and 5 are obtained from basis 2 by the successive addition of a $2p\sigma$ orbital on Li, a pair of $2p\pi$ orbitals on Li, and a set of $3d$ orbitals on F respectively.

little effect, but the further addition (basis 4) of lithium $p\pi$ orbitals allows a significant back-transfer of charge from the fluorine into the bond and onto the lithium, and provides a more balanced description of the $\sigma$–$\pi$ and ionic–covalent aspects of the bond. The main effect of the fluorine $d$ orbitals (basis 5) is to polarize electron density from behind the fluorine towards the lithium.

Despite its several limitations, Mulliken population analysis is currently widely used for interpretative purposes, mainly, presumably, because of the ease with which the populations can be calculated. As we have seen, the method is strongly dependent on the basis of atomic orbitals used to construct the wave function, and the Mulliken prescription for the calculation of the atomic charges may be criticized for its (arbitrary) equal division of the overlap populations between pairs of atoms. The basis-dependence can be avoided in general only by computing the atomic and bond charges directly from the charge-density function as in the method of Politzer and Harris (which can be extended to include bond charges by the definition of suitable bond regions; for example, Politzer and Reggio 1972). The discussions of tables 6.10 and 6.11 suggest however that studies of the basis-dependence of the charges can provide some detailed understanding of the changes in the electron distribution that accompany bond formation. The arbitrariness of the charges, whether arising from the different possible ways of partitioning the overlap populations or from different ways of defining atomic and bond regions in more direct methods, cannot however be avoided. It must therefore be concluded that no theoretical method of assigning charges to atoms and bonds can be entirely satisfactory. Mulliken population analysis is probably as good (or as bad) as any other method and, if used with some caution and with a proper regard for its limitations, can provide a valuable set of indices for the characterization of the electron distribution.

## 6.6 ORBITAL ENERGIES AND ATOMIC CHARGES

### 6.6.1 *Inner-shell binding energies*

We saw in §3.4 that, in accordance with Koopmans' theorem, orbital energies are often approximately observables in the sense that they provide estimates of the binding energies of electrons (vertical ionization energies) in atoms and molecules. These energies are accessible to the techniques of photoelectron spectroscopy, and the study of both valence and inner-shell binding energies in molecules has in recent years become a powerful tool for the determination and interpretation of molecular structure (Turner 1968; Siegbahn *et al.* 1969; Shirley 1973). We are concerned here primarily with the inner-shell electrons.

Unlike that of a valence electron, the charge distribution of an inner-shell electron is normally tightly localized about a single nucleus of a molecule, and remains relatively unchanged on going from the atom in the molecule to the free atom. The binding energy on the other hand can change significantly on going from the molecule to the free atom, or to another molecule. In table 6.12 are shown the 1s binding energies of carbon, nitrogen, oxygen and fluorine in a number of molecules, and the energy shifts relative to $CH_4$, $NH_3$, $H_2O$ and $CF_4$ respectively.† The values of the binding energies are seen to be well defined characteristic properties of the atoms, and their observation in photoelectron spectro-

TABLE 6.12  *1s binding energies and energy shifts (in units of* eV)

| Molecule | Carbon 1s | | Nitrogen 1s | | Oxygen 1s | | Fluorine 1s | |
|---|---|---|---|---|---|---|---|---|
| | Energy | Shift | Energy | Shift | Energy | Shift | Energy | Shift |
| $CH_4$ | 290.8 | (0.0) | | | | | | |
| $NH_3$ | | | 405.6 | (0.0) | | | | |
| $H_2O$ | | | | | 539.7 | (0.0) | | |
| $CF_4$ | 301.8 | +11.0 | | | | | 695.0 | (0.0) |
| $CHF_3$ | 299.1 | +8.3 | | | | | 694.1 | −0.9 |
| $CH_2F_2$ | 296.4 | +5.6 | | | | | 693.1 | −1.9 |
| $CH_3F$ | 293.6 | +2.8 | | | | | 692.4 | −2.6 |
| $C_6F_6$ | 294.0 | +3.2 | | | | | 693.7 | −1.3 |
| $C_6H_6$ | 290.4 | −0.4 | | | | | | |
| $C_2H_6$ | 290.6 | −0.2 | | | | | | |
| $C_2H_4$ | 290.7 | −0.1 | | | | | | |
| $C_2H_2$ | 291.2 | +0.4 | | | | | | |
| $O_2$ | | | | | 543.1 | +3.4 | | |
| CO | 296.2 | +5.4 | | | 542.6 | +2.9 | | |
| $CO_2$ | 297.6 | +6.8 | | | 541.1 | +1.4 | | |
| $CH_3OH$ | 292.7 | +1.9 | | | 538.9 | −0.8 | | |
| $(CH_3)_2CO$ | 291.2 | +0.4 | | | 539.0 | −0.7 | | |
| $(CH_3)_2CO$ | 293.8 | +3.0 | | | 539.0 | −0.7 | | |
| $SOF_2$ | | | | | 539.4 | −0.3 | 693.6 | −1.4 |
| $N_2$ | | | 410.0 | +4.4 | | | | |
| NNO | | | 408.8 | +3.2 | 541.2 | +1.5 | | |
| NNO | | | 412.6 | +7.0 | 541.2 | +1.5 | | |
| HCN | 293.4 | +2.6 | 406.2 | +0.6 | | | | |

Data from: Siegbahn *et al.* 1969; Davis *et al.* 1970; Davis, Shirley and Thomas 1972; Shirley 1973.

† The values of electron binding energies are conventionally quoted in units of eV.

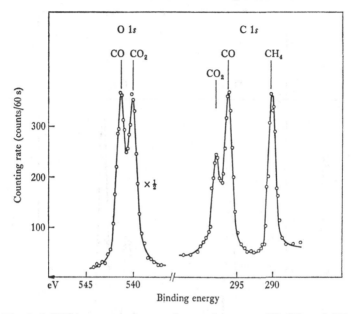

Fig. 6.16. ESCA spectrum from a mixture of the gases CO, $CO_2$ and $CH_4$ (Siegbahn *et al.* 1969, p. 13).

scopy provides a convenient method for the determination of the constitution of molecules. More interesting are the binding-energy shifts, which are measures of the local environments of the inner-shell electrons in different molecules or in different parts of the same molecule. Fig. 6.16 shows the shifts in the 1*s* spectrum of a mixture of CO, $CO_2$ and $CH_4$, whilst fig. 6.17 shows the carbon 1*s* shift associated with the presence of two types of carbon atom in acetone. The intensity ratio of the carbon peaks is the expected 1:2. The study of inner-shell shifts by X-ray photoelectron spectroscopy is called ESCA, for 'electron spectroscopy for chemical analysis'.

The use of Koopman's theorem (or an appropriate 'frozen-orbitals' method when the theorem does not apply) gives binding energies that are generally larger than the experimental values, particularly for inner-shell electrons (see table 4.2 for $CH_4$)†. Much of this difference can normally be accounted for by performing comparable Hartree–Fock

† There is no variation principle for orbital energies, so that approximate SCF calculations can give orbital energies that are either larger or smaller than the accurate Hartree–Fock values. In practice, the approximate orbital energies are nevertheless nearly always larger than the experimental binding energies.

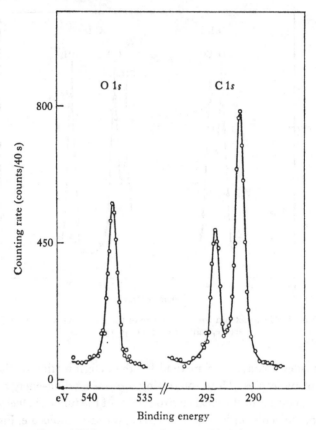

Fig. 6.17. 1*s* lines of ESCA spectrum of acetone (Siegbahn *et al.* 1969, p. 115).

calculations for the neutral and ionized species, in order to allow for the relaxation of the charge distribution in the presence of the positive hole in the ion. The relaxation energies associated with the removal of a 1*s* electron from C in $CH_4$, N in $NH_3$ and O in $H_2O$ are about $-14$, $-17$ and $-19$ eV respectively. The corresponding values in many other molecules differ from these by less than 2 eV (Aarons *et al.* 1973), so that the inner-shell binding-energy *shifts* predicted by the use of Koopmans' theorem can then be expected to be accurate to within approximately this quantity (2 eV). This is demonstrated by the plot in fig. 6.18 of the 1*s* orbital energy against the observed binding energy for carbon in a number of molecules. The orbital energies have all been calculated in terms of the same double-zeta contracted Gaussian basis of (10*s*, 5*p*/4*s*, 2*p*) for C, N, O and F, and (4*s*/2*s*) for H (Snyder and Basch 1972), and this use of a con-

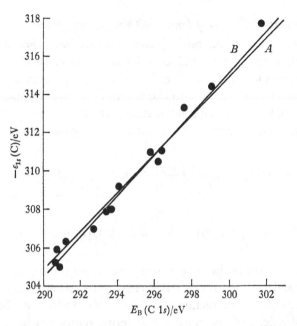

Fig. 6.18. Plot of the $1s$ orbital energy ($\epsilon_{1s}$) of carbon against the binding energy ($E_B$) for, in increasing value of $-\epsilon_{1s}$, $CH_4$, $C_2H_6$, $C_2H_4$, $C_2H_2$, $CH_3OH$, HCN, $CH_3F$, $H_2CO$, CO, HCOOH, $CH_2F_2$, $CO_2$, $CHF_3$ and $CF_4$. The orbital energies are double-zeta values taken from Snyder and Basch (1972). Line $A$ is the (least-squares) fit of unit slope, $-\epsilon_{1s} = E_B + 14.9$ eV, and line $B$ is the best fit, $-\epsilon_{1s} = 1.05\,E_B$.

sistent basis obviates to a great extent the effects of the basis-dependence of the energy shifts. The two lines drawn through the points in fig. 6.18 are $A$, a least-squares fit of unit slope, $-\epsilon_{1s} = E_B + 14.9$ eV, and $B$, a best fit $-\epsilon_{1s}/E_B = 1.05$, where $\epsilon_{1s}$ is the carbon $1s$ orbital energy and $E_B$ is the binding energy. The latter simply shows that the magnitudes of the orbital energies are on average about 5 per cent larger than the binding energies, and gives a maximum error (for $CO_2$ in fig. 6.18) of 0.8 eV. Line $A$ shows that the average contribution of relaxation, correlation and relativistic effects to the carbon $1s$ binding energy is $-14.9$ eV. All but about 1.2 eV of this is due to the relaxation effect. The maximum deviation of the points from line $A$ is 1.0 eV (for $CF_4$), and this is a measure both of the variation of the relaxation energy amongst the molecules and of the possible errors in the binding-energy shifts predicted from orbital-energy differences. Similar results are obtained for the inner-shell shifts of other atoms.

## The electron distribution

### 6.6.2 The potential model of inner-shell binding-energy shifts

The origin of inner-shell binding-energy shifts lies primarily in the different electrostatic potential energies of an inner-shell electron in different molecules. We consider, for example, an electron in orbital $1s_A$ (a localized inner-shell molecular orbital†) centred on nucleus A of a molecule in a closed-shell ground state (Schwartz 1970; Basch 1970). The orbital energy (3.27) of this electron can be written as (in units of $H_\infty$)

$$\epsilon(1s_A) = \epsilon^0(1s_A) - \sum_{\alpha \neq A} Z_\alpha \int \frac{1s_A(\boldsymbol{r})^2}{r_\alpha}\,\mathrm{d}v + \sum_{n \neq 1s_A} (2J_{n,1s_A} - K_{n,1s_A})$$

(6.43)

where $\quad \epsilon^0(1s_A) = \int 1s_A(\boldsymbol{r})\left[-\tfrac{1}{2}\nabla^2 - Z_A/r_A\right] 1s_A(\boldsymbol{r})\,\mathrm{d}v + J_{1s_A\,1s_A}$

is the kinetic energy of the $1s_A$ electron plus its energy of interaction with nucleus A and with the other $1s_A$ electron. The second set of terms in (6.43) represents the interaction of the $1s_A$ electron with the other nuclei in the molecule, and the third set of terms represents its Coulomb-exchange interaction with the electrons in other orbitals.

Because of the highly localized nature of the inner shell, the exchange terms $K_{n,1s_A}$ are small compared with the other terms in (6.43). An approximate expression for the orbital energy is therefore

$$\epsilon(1s_A) \simeq \epsilon^0(1s_A) + V(1s_A)$$

(6.44)

where $\quad V(1s_A) = - \sum_{\alpha \neq A} Z_\alpha \int \frac{1s_A(\boldsymbol{r})^2}{r_\alpha}\,\mathrm{d}v + \int \frac{1s_A(\boldsymbol{r}_1)^2 P'(\boldsymbol{r}_2)}{r_{12}}\,\mathrm{d}v_1\,\mathrm{d}v_2 \quad$ (6.45)

and $\qquad P'(\boldsymbol{r}) = 2 \sum_{n \neq 1s_A} |\psi_n(\boldsymbol{r})|^2 = P(\boldsymbol{r}) - 2[1s_A(\boldsymbol{r})^2]$

is the molecular charge density function excluding the $1s$ electrons on A. The quantity $V(1s_A)$ is the electrostatic potential energy of an electron in $1s_A$ in the presence of the charges outside the inner shell. Because the 'one-centre' quantity $\epsilon^0(1s_A)$ can be expected to be insensitive to changes in the molecule outside the inner shell, it can be regarded as a constant

† Although the inner-shell molecular orbitals in, for example, $O_2$ are two-centre orbitals ($1\sigma_g$ and $1\sigma_u$), Bagus and Schaefer (1972) have shown that the observed inner-shell binding energies can be understood only if the ionization process involves the removal of an electron from the vicinity of *one* nucleus only; that is, if the inner-shell orbitals are localized. This is true generally for inner shells, but not for valence shells.

characteristic of the inner shell. A change in the orbital energy due to a change in the environment of the inner shell is then given by

$$\Delta\epsilon(\text{is}_A) \simeq \Delta V(\text{is}_A) \tag{6.46}$$

The expression (6.45) for the potential energy can be simplified further if the $\text{is}_A$ orbital is collapsed into nucleus A; that is, if the charge distribution of the inner-shell electron is approximated as a point charge on the nucleus. Equation (6.45) then becomes

$$V(\text{is}_A) \simeq - \sum_{\alpha \neq A} \frac{Z_\alpha}{R_{A\alpha}} + \int \frac{P'(\mathbf{r})}{r_A} \, dv \tag{6.47}$$

and $- V(\text{is}_A)$, in units of $H_\infty/e$, is the electrostatic potential at nucleus A due to the charges outside the inner shell on A. The use of (6.47) in (6.46) is demonstrated in fig. 6.19 for the $\text{is}$ shifts of carbon, nitrogen, oxygen and fluorine in a number of molecules (Schwartz 1970). The average value of $\Delta\epsilon/\Delta V$ is 1.1, and the maximum deviation from the line $\Delta\epsilon = \Delta V - 0.28\,\text{eV}$ is $0.6\,\text{eV}$, for carbon in CO.

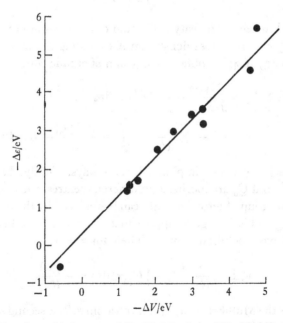

Fig. 6.19. Plot of $\text{is}$ orbital-energy shifts (C shifts from $CH_4$, N shifts from $NH_3$, O shifts from $H_2O$, and F shifts from HF) against potential-energy shifts for, in order of increasing value of $-\Delta\epsilon$, $CH_3F$, $H_2CO$, $C_2H_2$, FOH, HCN, HCN, $CH_3F$, CO, FOH, $H_2CO$ and CO (Schwartz 1970).

## The electron distribution

### 6.6.3 Correlation with atomic charges

The charge density function $P'(r)$ in (6.47) can be partitioned into contributions associated with the different atoms in a molecule by the methods discussed in §6.5, and such a partitioning can then be used to express both the potential energy (6.47) and the shift (6.46) in terms of simple molecular parameters. Consider, for example, the partitioning (6.34) of the total molecular density function:

$$P(r) = \sum_\alpha P_\alpha(r) + \sum_{\alpha > \beta} \sum P_{\alpha\beta}(r)$$

By analogy with the definition (6.40) of the gross atomic populations, we can define a set of gross atomic density functions,

$$Q_\alpha(r) = P_\alpha(r) + \tfrac{1}{2} \sum_{\beta \neq \alpha} P_{\alpha\beta}(r)$$

such that the gross populations are

$$N(\alpha) = \int Q_\alpha(r) \, dv$$

and the total molecular density is the sum of the gross atomic densities. Then, if $Q'_A(r)$ is the gross density on A excluding the inner shell, the potential energy (6.47) is obtained as a sum of atomic contributions:

$$V \simeq \sum_\alpha V_\alpha = \int \frac{1 s_A(r_1)^2 \, Q'_A(r_2)}{r_{12}} \, dv_1 \, dv_2$$
$$- \sum_{\alpha \neq A} \left\{ \frac{Z_\alpha}{R_{A\alpha}} - \int \frac{1 s_A(r_1)^2 \, Q_\alpha(r_2)}{r_{12}} \, dv_1 \, dv_2 \right\} \quad (6.48)$$

This expression can be simplified in two ways. Firstly, because the densities $1 s_A^2$ and $Q_\alpha$ are localized on different centres, the denominator $r_{12}$ in the last group of terms in (6.48) can be replaced by the internuclear distance $R_{A\alpha}$. This is a good approximation except, possibly, when a direct bond exists between A and $\alpha$. Then, for $\alpha \neq A$,

$$V_\alpha \simeq - \frac{1}{R_{A\alpha}} \left\{ Z_\alpha - \int Q_\alpha(r) \, dv \right\} = - \frac{q(\alpha)}{R_{A\alpha}}$$

where $q(\alpha)$ is the Mulliken charge (6.42) on atom $\alpha$. The second simplification is obtained by setting $\quad V_A \simeq l - kq(A)$

where $k$ and $l$ are assumed to be (positive) constants characteristic of the

166

inner shell. This can be justified on the grounds that a change in $V_A$ is due primarily to a change in the population of the valence orbitals on A, with $V_A$ increasing (becoming more positive) as the population increases. The constant $k$ is then interpreted as an average potential for the inter-action of a $1s_A$ electron with a valence electron on A. The total potential energy (6.48) now becomes

$$V \simeq -\{kq(A) + \sum_{\alpha \neq A} q(\alpha)/R_{A\alpha} - l\}$$

and the orbital-energy shift (6.45) is given by

$$-\Delta\epsilon \simeq k\Delta q(A) + \Delta\{\sum_{\alpha \neq A} q(\alpha)/R_{A\alpha}\}$$

or, equivalently,

$$-\Delta\epsilon \simeq kq(A) + \sum_{\alpha \neq A} q(\alpha)/R_{A\alpha} + C \qquad (6.49)$$

where $C$ is an appropriate reference energy. The validity of this type of relation (using Mulliken charges) is demonstrated in fig. 6.20 for the carbon inner shell (Shirley 1973).

Although the potential $k$ in this point-charge model of orbital shifts can be estimated from suitable calculations on atoms (Gelius, Roos and Siegbahn 1970; Politzer and Politzer 1973), (6.49) is usually, and more reliably, treated as an empirical relation. The constants $k$ and $C$ are then

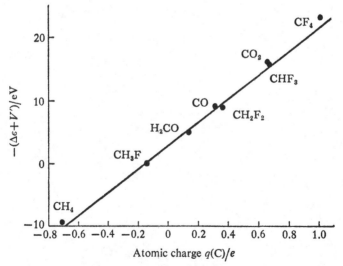

Fig. 6.20. Point-charge model of orbital shifts $(-\Delta\epsilon = kq(C) + V' + C$, equation (6.49)) for the carbon inner shell (Shirley 1973).

obtained by fitting computed atomic charges either to computed orbital energies or to experimental binding energies. One advantage of this empirical use of (6.49) is that the dependence of the charges on the method of computation and on the basis is thereby largely avoided, different sets of charges giving very similar fits, although with different values of the constants in (6.49). It also allows the use of charges obtained from semi-empirical theories (Siegbahn *et al.* 1969; Shirley 1973).

# 7 The chemical bond

The interaction of two atoms can lead to the formation of a stable chemical bond if the total energy of the system of interacting atoms is lower than that of the separated atoms. An understanding of the nature of the chemical bond therefore requires an explanation of how the redistribution of electronic charge that accompanies bond formation leads to a lowering of the energy. For example, it has long been recognized that the constructive interference of atomic wave functions gives rise to an accumulation of charge between the nuclei, and the attractive forces between this charge and the nuclei have often been held responsible for the stability of the covalent bond. The accumulation of charge in the binding region between the nuclei cannot however by itself provide a satisfactory explanation of the stability of the bond since it involves a transfer of electronic charge from regions of low potential energy near the nuclei to a region of higher potential energy. In addition, the example of $N_2$ (fig. 6.3) shows that charge is often also transferred from the regions near the nuclei into the antibinding regions behind the nuclei, and this not only results in an increase of the potential energy of the electrons, but also accentuates the repulsion between the nuclei. On the other hand, as we shall see, the removal of electronic charge from the regions near the nuclei results in a decrease of the kinetic energy of the electrons. An understanding of the energy changes that accompany bond formation therefore requires an understanding of the relationship between the total, potential and kinetic energies. This relationship is provided by the virial theorem.

## 7.1 THE VIRIAL THEOREM

Consider the Hamiltonian (1.2) of a system of electrons and nuclei in which the charges are assumed to interact through electrostatic forces only. The virial theorem for a stationary state of such a system of charges is

$$2\langle T \rangle + \langle V \rangle = 0 \tag{7.1}$$

where $\langle T \rangle$ and $\langle V \rangle$ are the expectation values of the kinetic and potential energies respectively (for a good discussion, see Löwdin 1959b). The total energy is

$$E = \langle T \rangle + \langle V \rangle$$

*The chemical bond*

and, therefore,
$$\langle T \rangle = -E, \quad \langle V \rangle = 2E \tag{7.2}$$

In the Born–Oppenheimer approximation, the Hamiltonian for the electronic states of a molecule is given by (1.4), without the kinetic-energy terms for the nuclei and with the nuclei assumed to be stationary. The theorem, as given by (7.1), is then generally valid only for an equilibrium configuration of the molecule and for the separate atoms. Thus, for a diatomic molecule, (7.1) is replaced by

$$2\langle T \rangle + \langle V \rangle + R\frac{dE}{dR} = 0 \tag{7.3}$$

where $R$ is the internuclear separation, and (7.2) becomes

$$\langle T \rangle = -E - R\frac{dE}{dR}, \quad \langle V \rangle = 2E + R\frac{dE}{dR} \tag{7.4}$$

Whenever $dE/dR = 0$, as at the equilibrium distance $R_e$ and at infinite separation of the nuclei, (7.3) reduces to the simpler form (7.1).

The importance of the virial theorem in a discussion of the energy changes that accompany bond formation is that it provides a necessary relationship between the changes of kinetic and potential energies. If $\Delta E$, $\Delta T$ and $\Delta V$ are the differences, respectively, between the total, kinetic and potential energies of a diatomic molecule and of the separate atoms then, at any distance $R$,

$$2\Delta T + \Delta V + R\frac{dE}{dR} = 0 \tag{7.5}$$

$$\Delta T = -\Delta E - R\frac{dE}{dR}, \quad \Delta V = 2\Delta E + R\frac{dE}{dR} \tag{7.6}$$

At the equilibrium distance,

$$\Delta_e T = -\tfrac{1}{2}\Delta_e V = -\Delta_e E = D_e \tag{7.7}$$

where $D_e$ is the dissociation energy of the molecule. The formation of a stable bond therefore necessarily results in an *increase* of the kinetic energy and a *decrease* of the potential energy.

The virial theorem is satisfied by exact eigenfunctions of the Hamiltonian and by certain classes of approximate wave functions, including accurate Hartree–Fock wave functions. An arbitrary approximate wave function can however always be made to satisfy the theorem by means of a variational scaling procedure (Löwdin 1959$b$). Consider for example the ground state of a helium-like atom, with nuclear charge $Ze$, for which a simple (spinless) orbital wave function is

$$\Psi(r_1, r_2) = 1s(r_1)\, 1s(r_2)$$

where
$$1s(r) = (\zeta^3/\pi)^{\frac{1}{2}}\, e^{-\zeta r}$$

is the ground-state orbital of a hydrogen-like atom with ('effective') nuclear charge $\zeta e$. The corresponding energy is a function of $\zeta$,

$$E(\zeta) = \zeta^2 - 2Z\zeta + \tfrac{5}{8}\zeta$$

(in units of $H_\infty$), and the kinetic and potential energies are

$$\langle T \rangle = \zeta^2, \quad \langle V \rangle = -\zeta(2Z - \tfrac{5}{8})$$

The parameter $\zeta$ acts as a scale factor for the wave function. Thus, an increase in the value of $\zeta$ results in a contraction of the wave function towards the nucleus, and this contraction is accompanied by an increase of the kinetic energy of the electrons and a decrease of the potential energy, due to the greater localization of the electrons in the vicinity of the nucleus. The virial theorem is satisfied for only one value of $\zeta$:

$$\langle T \rangle = -\tfrac{1}{2}\langle V \rangle \quad \text{when} \quad \zeta = Z - \tfrac{5}{16}$$

and this is just the value which is obtained by the variation principle. Thus,

$$\frac{\partial E}{\partial \zeta} = 2\zeta - 2Z + \tfrac{5}{8}$$

$$= 0 \quad \text{when} \quad \zeta = Z - \tfrac{5}{16}$$

The minimization of the energy with respect to a scale parameter therefore produces a wave function which automatically satisfies the virial theorem. This is a general result for atoms and molecules.

In the following section we discuss how the redistribution of charge which leads to the energy changes (7.7) can be interpreted to a large extent in terms of two effects: delocalization (including polarization) and contraction.

## 7.2 DELOCALIZATION AND CONTRACTION

### 7.2.1 *The ground state of* $H_2^+$

Ruedenberg and co-workers (Feinberg, Ruedenberg, and Mehler 1970; Feinberg and Ruedenberg 1971) have analysed the energy changes that accompany the formation of the bond in the ground state of $H_2^+$ by considering the behaviour of three approximate wave functions of the form

$$\psi(\mathbf{r}) = N[A(\mathbf{r}) + B(\mathbf{r})] \tag{7.8}$$

where $N$ is a normalization factor, and $A(\mathbf{r})$ and $B(\mathbf{r})$ are equivalent 'atomic orbitals' associated with the two nuclear centres (labelled A and B). Such a wave function can be the exact ground-state wave function if $A$ and $B$ are suitably chosen. The approximate wave functions considered by Feinberg and Ruedenberg (1971) are:

(1) The simple LCAO–MO function due to Pauling (1928), in which $A$ and $B$ are ground-state orbitals of the free hydrogen atom:

$$\left. \begin{aligned} A = 1s(r_A) = \pi^{-\frac{1}{2}} e^{-r_A}, \quad B = 1s(r_B) = \pi^{-\frac{1}{2}} e^{-r_B} \\ \psi_P(\mathbf{r}) = N[1s(r_A) + 1s(r_B)] \end{aligned} \right\} \tag{7.9}$$

(2) The Finkelstein–Horowitz (1928) wave function $\psi_{FH}$ obtained by including a scale parameter, or 'effective nuclear charge', $\zeta$ in the atomic orbitals:

$$A = (\zeta^3/\pi)^{\frac{1}{2}} e^{-\zeta r_A}, \quad B = (\zeta^3/\pi)^{\frac{1}{2}} e^{-\zeta r_B} \tag{7.10}$$

The parameter $\zeta$ is a function of the internuclear distance $R$, and is determined by minimization of the energy at each value of $R$.

(3) The Guillemin–Zener (1929) wave function $\psi_{GZ}$ obtained by choosing

$$A = C e^{-(\zeta_1 r_A + \zeta_2 r_B)}, \quad B = C e^{-(\zeta_1 r_B + \zeta_2 r_A)} \tag{7.11}$$

where $C$ is a normalization factor, and $\zeta_1$ and $\zeta_2$ are again determined by minimization of the energy at each value of $R$.

The equilibrium distance $R_e$, total energy $E(R_e)$, and dissociation energy $D_e$ obtained with each of these wave functions are compared with the exact values in table 7.1. For the present purpose, the Guillemin–Zener wave function can be considered as exact.

TABLE 7.1  *The ground state of* $H_2^+$

| Wave function | $R_e/a_\infty$ | $-E(R_e)/H_\infty$ | $D_e/H_\infty$ |
|---|---|---|---|
| Pauling | 2.50 | 0.565 | 0.065 |
| Finkelstein–Horowitz ($\zeta = 1.24$) | 2.00 | 0.587 | 0.087 |
| Guillemin–Zener ($\zeta_1 = 1.14$, $\zeta_2 = 0.22$) | 2.00 | 0.6024 | 0.1024 |
| Exact | 2.00 | 0.6026 | 0.1026 |

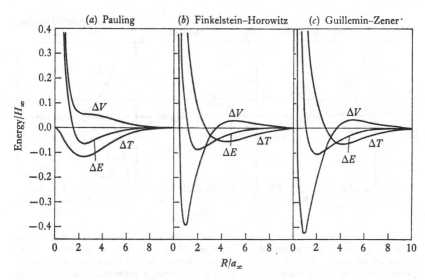

Fig. 7.1. The ground state of $H_2^+$. Dependence on the internuclear distance $R$ of the total binding energy $\Delta E$, and of its kinetic-energy and potential-energy components, $\Delta T$ and $\Delta V$.

In fig. 7.1 are shown the dependence on the internuclear distance of the total binding energy, $\Delta E(R) = E(R) - E(\infty)$, and of its kinetic-energy and potential-energy components, $\Delta T(R)$ and $\Delta V(R)$, for the three approximate wave functions. It is clear that the Finkelstein–Horowitz wave function gives the correct behaviour of the $\Delta E$, $\Delta T$ and $\Delta V$ curves over the range of values of $R$ of interest. The Pauling wave function, on the other hand, gives a qualitatively correct $\Delta E$ curve, but *incorrect* $\Delta T$ and $\Delta V$ curves. An understanding of the bonding can therefore be obtained from a comparison of the Pauling and Finkelstein–Horowitz approximations.

The ground-state binding energy of $H_2^+$ can be written as

$$\Delta E(R) = E(R) - E(\infty) = \frac{1}{R} - \int \frac{1s(r_A)^2}{r_B}\, dv + \Delta E'(R) \qquad (7.12)$$

where

$$\frac{1}{R} - \int \frac{1s(r_A)^2}{r_B}\, dv = e^{-2R}(1 + R)/R = \Delta E^0(R)$$

is the potential energy of interaction of an unperturbed hydrogen atom with a proton, and $\Delta E'$ is the change in energy due to the reorganization of the electron distribution on bond formation. Fig. 7.2(*a*) shows that $\Delta E^0$ is negligible for internuclear distances greater than $3a_\infty$, and becomes

173

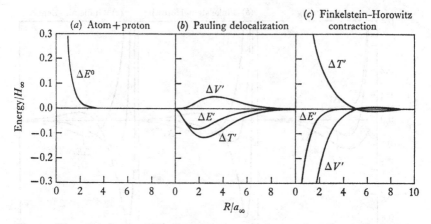

Fig. 7.2. The ground state of $H_2^+$. Dependence on the internuclear distance $R$ of ($a$) the potential energy of interaction $\Delta E^0$ of an unperturbed hydrogen atom with a proton, ($b$) the reorganization energy (or delocalization energy) $\Delta E'$, and its kinetic-energy and potential-energy components $\Delta T'$ and $\Delta V'$, in the Pauling approximation, and ($c$) the additional reorganization (or contraction) energy and its components in the Finkelstein–Horowitz approximation.

significant only at distances less than the equilibrium distance; thus,

$$\Delta E \simeq \Delta E' \quad \text{for} \quad R > R_e$$

The energy changes that accompany the approach to equilibrium can therefore be attributed almost wholly to the reorganization of the electron distribution.

The dependence on $R$ of the reorganization energy $\Delta E'$ in the Pauling approximation is shown in fig. 7.2($b$), and the *additional* reorganization energy in the Finkelstein–Horowitz approximation in fig. 7.2($c$). We consider first the long-range behaviour. For internuclear distances greater than about $5a_\infty$ the three approximate wave functions, $\psi_P$, $\psi_{FH}$, $\psi_{GZ}$, are almost identical, with the variation principle giving $\zeta \simeq 1$ for $\psi_{FH}$, and $\zeta_1 \simeq 1$, $\zeta_2 \simeq 0$ for $\psi_{GZ}$. The corresponding energy curves are therefore also almost identical, and it is sufficient to consider the Pauling approximation only. As the atoms approach each other from infinity, fig. 7.2($b$) shows that the potential energy increases but the kinetic energy decreases more strongly, with a resultant net decrease of the total energy. Thus, the constructive interference of the atomic wave functions leads to a transfer of charge from the regions of low potential energy near the nuclei towards the region of higher potential energy between the nuclei, and therefore results in an increase of the potential energy. On the other hand, this *delocalization* of the charge along the direction of the molecular axis results in a decrease of the magnitude of the momentum of the electron along this direction and, therefore, to a decrease of the corresponding component of the kinetic energy. The components perpendicular to the molecular axis remain relatively unchanged. The lowering of the total energy of the molecule at large internuclear distances is therefore essentially a delocalization effect.

For internuclear distances between $5a_\infty$ and $2a_\infty$, figs. 7.1 ($a$) and 7.2 ($b$) show that the interpretation of the bonding remains unchanged in the Pauling approximation, with $\Delta V$ positive and $\Delta T$ negative at the equilibrium distance, in contradiction to the requirement (7.7) of the virial theorem. The two more accurate wave functions on the other hand satisfy the virial theorem at all values of $R$. The Finkelstein–Horowitz wave function is a scaled Pauling wave function, with the scale parameter $\zeta$ determined by the variation principle. The value of $\zeta$ is close to unity when $R > 5a_\infty$, but increases as the nuclei come closer together to a value of 1.24 at $R_e = 2a_\infty$ (at $R = 0$, $\zeta = 2$ and $\psi_{FH}$ is the exact ground-state wave function of the united atom $He^+$). The increase in the value of $\zeta$ corresponds to a contraction of the wave function with the nuclei as contraction centres, and is accompanied by an increase of the kinetic energy and a (greater) decrease of the potential energy (fig. 7.2 ($c$)). This contraction of the wave function may be envisaged (rather loosely) as a consequence of the delocalization effect, which removes electronic charge from the vicinity of the nuclei. The remaining atom-like charge distribution near each nucleus experiences an enhanced attraction towards the nucleus, and the resulting one-centre contraction at each nucleus is the major cause of the energy changes. The charge distribution between the nuclei, on the other hand, is attracted towards both nuclei, and is therefore contracted towards the internuclear axis. The charge in the outer regions of the molecule is contracted towards the centre of the molecule.

In summary, the comparison of the Pauling and Finkelstein–Horowitz approximations shows that the reorganization of the electron distribution that accompanies the formation of the bond in $H_2^+$ can be described to a large extent as the combination of two effects: delocalization along the direction of the internuclear axis, and contraction towards the nuclei as contraction centres. The corresponding changes in the charge distribution are shown in fig. 7.3. The delocalization distribution in fig. 7.3($a$) is the differences, at $R = 2a_\infty$, between the Pauling density $P_P = \psi_P^2$ and the symmetrized density of the system of proton plus unperturbed

## The chemical bond

hydrogen atom:

$$\Delta P_{\text{deloc}}(\boldsymbol{r}) = P_{\text{P}}(\boldsymbol{r}) - \tfrac{1}{2}[1s(r_{\text{A}})^2 + 1s(r_{\text{B}})^2]$$

(the symmetrization of the unperturbed distribution can itself be regarded as an additional delocalization, which is present in $H_2^+$ and other

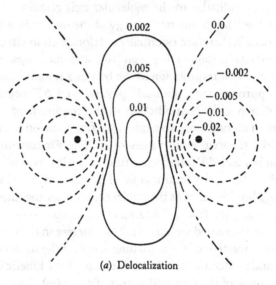

(a) Delocalization

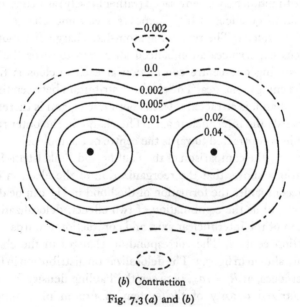

(b) Contraction

Fig. 7.3 (a) and (b)

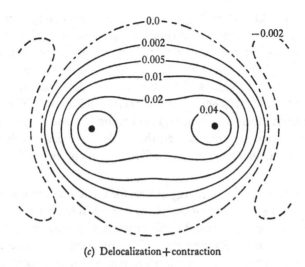

(c) Delocalization + contraction

Fig. 7.3. The ground state of $H_2^+$. Reorganization of the electron distribution at $R = 2a_\infty$: (a) the delocalization distribution $\Delta P_{deloc}$, (b) the contraction distribution $\Delta P_{contr}$, and (c) delocalization plus contraction.

symmetrical molecular ions, but not normally in neutral molecules). The contraction distribution shown in fig. 7.3(b) is the difference, again at $R = 2a_\infty$, between the Finkelstein–Horowitz density $P_{FH} = \psi_{FH}^2$ and the Pauling density:

$$\Delta P_{contr}(\mathbf{r}) = P_{FH}(\mathbf{r}) - P_P(\mathbf{r})$$

The two effects are combined in fig. 7.3(c):

$$\Delta P(\mathbf{r}) = \Delta P_{deloc}(\mathbf{r}) + \Delta P_{contr}(\mathbf{r}) = P_{FH}(\mathbf{r}) - \tfrac{1}{2}[1s(r_A)^2 + 1s(r_B)^2]$$

Although the Finkelstein–Horowitz approximation gives an essentially correct description of the bonding in $H_2^+$, it lacks one important feature. We saw in chapter 5 that a satisfactory description of bonding in molecules can in general be obtained within the LCAO–MO scheme only if the basis includes polarization functions to describe the distortion of the atomic orbitals in the molecular environment. The Finkelstein–Horowitz wave function is a combination of spherical atomic orbitals, and the absence of atomic polarization accounts for the deficiencies of the approximation. On the other hand, the 'atomic orbitals' (7.11) used to construct the Guillemin–Zener wave function are not spherical, the function $A$, for example, containing a factor which is centred on B. The inclusion of this factor causes a polarization of the atomic orbital on A towards B, and allows a more accurate representation of the redistribution of charge parallel to the bond direction. The corresponding

## The chemical bond

charge-density difference, $P_{GZ}(\boldsymbol{r}) - P_{FH}(\boldsymbol{r})$, is qualitatively similar to the (interference) delocalization distribution shown in fig. 7.3(a), and atomic polarization can therefore be conveniently classed as one aspect of the delocalization effect. Unlike the delocalization due to the interference of the atomic wave functions, however, atomic polarization leads to a decrease of the potential energy of the electron since it involves a transfer of charge from the antibinding regions behind the nuclei into the region of lower potential energy between the nuclei, whereas interference removes charge mainly from the immediate vicinity of the nuclei.

*The first excited state.* The energy curves for the lowest antibonding state of $H_2^+$ show a behaviour quite opposite to that in fig. 7.1 for the ground state (Feinberg *et al.* 1970). The destructive interference of the atomic wave functions in the Pauling approximation results in a sharp *increase* of the kinetic energy of the electron and a smaller decrease of the potential energy (table 7.2), corresponding to a movement of charge from the centre of the molecule onto the nuclei and into the antibinding regions. The introduction of a scale parameter in the Finkelstein–Horowitz approximation then results in an *expansion* of the wave function, which is accompanied by a decrease of the kinetic energy and an almost equal increase of the potential energy. The Guillemin–Zener approximation shows very little additional change, so that atomic polarization is not important in this case. It is clear from table 7.2 that the increase of kinetic energy due to interference dominates the energy changes.

TABLE 7.2 *Energy changes at $R = 2a_\infty$ for the first excited state of $H_2^+$*

| Wavefunction | $\Delta E/H_\infty$ | $\Delta T/H_\infty$ | $\Delta V/H_\infty$ |
|---|---|---|---|
| Pauling | 0.3391 | 0.4363 | −0.0972 |
| Finkelstein–Horowitz ($\zeta = 0.90$) | 0.3342 | 0.3082 | 0.0260 |
| Guillemin–Zener ($\zeta_1 = 0.90$, $\zeta_2 = -0.002$) | 0.3342 | | |

### 7.2.2 The ground state of $He_2$

Whereas the bonding in the ground state of $H_2$ is similar to that in $H_2^+$, with electron interaction causing only minor modifications, the effects of electron interaction in $He_2$ are large enough to make the ground state

178

of the molecule unstable with respect to the separated ground-state atoms. Two effects can be distinguished in $He_2$: Coulomb repulsion between the electrons, and 'exchange repulsion' between electrons with the same spin. The former is an essentially classical effect which, as in $H_2$, is not sufficiently strong to prevent the formation of a stable bond. Exchange repulsion, on the other hand, is a consequence of the Pauli (antisymmetry) principle, and prevents electrons with the same spin from occupying the same region of space. In $He_2$ it results in a net antibonding interaction between the two pairs of electrons, with a transfer of charge out of the region between the nuclei onto the nuclei and into the antibinding regions, as shown by the density-difference distribution in fig. 7.4 (*a*). The corresponding distribution *in the absence of the Pauli principle* would be similar to that of $H_2$ shown in fig. 7.4 (*b*).

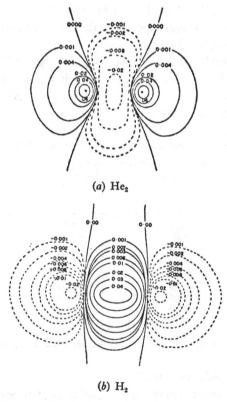

(*a*) $He_2$

(*b*) $H_2$

Fig. 7.4. Density-difference distributions at $R = 2a_\infty$ in the ground states of (*a*) $He_2$ and (*b*) $H_2$ (Bader and Chandra 1968; reproduced by permission of the National Research Council of Canada).

## The chemical bond

Some understanding of the relative strengths of the Coulomb and exchange interactions of electrons is obtained by comparing the results of orbital calculations with and without inclusion of the Pauli principle. For an $N$-electron atom the ground-state orbital configuration in the absence of the Pauli principle is $(1s)^N$, with all the electrons occupying the $1s$ orbital. The corresponding wave function is the product

$$\Psi(1, 2, ..., N) = 1s(1)\, 1s(2)\, 1s(3) ... 1s(N)$$

and the energy is
$$E = Nf_{1s} + \tfrac{1}{2}N(N-1)J_{1s,1s}$$

The corresponding results shown in table 7.3 are for the hydrogen-like orbital
$$1s(r) = (\zeta^3/\pi)^{\frac{1}{2}} e^{-\zeta r}$$

with set (i) for $\zeta = Z$, the atomic number, and set (ii) for the optimum value of $\zeta = Z - 5(N-1)/16$. Except for He, all the energies are lower than the Hartree–Fock (and experimental) values. The effect of the Pauli principle is to restrict the occupation of the orbitals to two electrons per orbital, thereby lowering the total kinetic energy and Coulomb interaction energy of the electrons, but decreasing the magnitude of the nuclear attraction energy. The net result is an increased total energy, despite the decrease of the Coulomb interaction energy.

TABLE 7.3 *Ground-state energies of atoms (in units of $H_\infty$)*

| Atom | Without the Pauli principle | | With the Pauli principle |
| | (i) | (ii) | RHF |
| --- | --- | --- | --- |
| He | −2.75 | −2.85 | −2.86 |
| Li | −7.88 | −8.46 | −7.43 |
| Be | −17.00 | −18.76 | −14.57 |
| B | −31.25 | −35.16 | −24.53 |
| C | −51.75 | −59.07 | −37.69 |
| N | −79.63 | −91.93 | −54.40 |
| O | −116.00 | −135.14 | −74.81 |
| F | −162.00 | −190.13 | −99.41 |
| Ne | −218.75 | −258.30 | −128.55 |
| Ar | −1194.75 | −1448.75 | −526.82 |

### 7.2.3 Covalent and ionic bonding
Although the interpretation of the chemical bond in molecules with more than one electron is complicated both by the presence of electron interaction and by the requirements of the Pauli principle, the analysis of

molecular bonding in terms of delocalization and contraction effects appears to be generally valid (Ruedenberg 1962; Wilson and Goddard 1972). Thus, within the orbital (RHF) approximation, the reorganization of the electron distribution can be interpreted in terms of (*a*) constructive interference, polarization and contraction for the bonding orbitals, and (*b*) destructive interference, polarization and expansion for the antibonding orbitals. The formation of a stable bond then involves an overall delocalization of the charge distribution along the direction of the bond, and an overall contraction towards the nuclei. The contraction effect provides the decrease of potential energy required by the virial theorem, but at the expense of an excessive increase of the kinetic energy. The delocalization effect, on the other hand, causes a large decrease of the kinetic energy, which compensates for the excessive increase on contraction and which, therefore, makes the contraction possible.

This interpretation of the chemical bond is consistent with the (RHF) density-difference distribution for $Li_2$ and $N_2$ shown in fig. 6.3. In $Li_2$, the absence of valence antibonding orbitals results in a mainly one-way transfer of charge into the region between the nuclei, which is similar to that found in $H_2^+$ and $H_2$ except that the atomic polarization (of the inner shells) is in the opposite sense. In $N_2$ the occupation of both bonding and antibonding orbitals, which is a consequence of the Pauli principle, causes a marked two-way transfer of charge which is accompanied by a stronger contraction of the charge distribution between the nuclei. The relative diffuseness of the $\Delta P$ distribution in the antibinding regions is then explained by the weaker contraction of the bonding orbitals in these regions and by the expansion of the $(2\sigma_u)$ antibonding orbital. The two-way transfer of charge and the $p\pi$-type negative $\Delta P$ distribution on each atom are largely due to a strong atomic polarization effect in the valence shell, which can be interpreted in terms of the mixing or hybridization of the $2s$ and $2p\sigma$ orbitals on each atom:

$$(2s)^2(2p\sigma) \rightarrow (h_1)^2(h_2)$$

where
$$h_i = N_i(2s + C_i 2p\sigma)$$

with $h_1$ pointing out of the binding region and $h_2$ into the binding region. Unlike the atomic polarization in $H_2^+$, which causes a transfer of charge from the antibinding regions onto the nuclei and into the binding region and which results in a lowering of the potential energy, this additional 'valence hybridization' causes a two-way transfer of charge away from the nuclei, and results in a lowering of the kinetic energy, thereby enhancing the effect of interference.

Whereas the covalent bond is characterized by a symmetrical accumulation of charge between the bonded atoms, we saw in §6.3 that the bonding in the highly ionic LiF is conveniently described in terms of the transfer of an electron from the lithium onto the more electronegative fluorine, with some back-polarization towards the lithium. A complete transfer of one electron would give a dipole moment of 7.51 D (at the equilibrium distance) compared with the observed value of 6.28 D, and the ratio of these gives an effective 'ionic character' of 84 per cent, corresponding to the transfer of 0.84 electrons (Bader and Henneker 1965). A similar result is obtained for LiH, with an effective transfer of 0.78 electrons from Li to H. The direction of the dipole moment is not however always consistent with a transfer of electronic charge towards the more electronegative atom. The classic example is CO, for which the small dipole moment of 0.1 D corresponds to $C^-O^+$. A more striking example is BF, whose dipole moment of 1.0 D has polarity $B^-F^+$ (Huo 1965; Bader and Bandrauk 1968).

The reorganization of the electron distribution which accompanies the formation of a polar bond can be discussed, at least in part, in terms of the relative delocalization-contraction characteristics of the bonded atoms. This type of discussion is then closely related to the more traditional approach which makes use of the empirical concept of electronegativity (Pauling 1960), the electronegativity of an atom being regarded as a measure of the latent delocalization and contraction of the atomic electron distribution, and of the associated changes of kinetic and potential energies. We consider first the bonding in LiF. We have seen (§ 6.3) that whereas the delocalization effect in $Li_2$ involves a mainly one-way transfer of charge into the region between the nuclei, that in $F_2$ involves a two-way transfer of charge, with the larger part (associated with the lone pairs) going into the antibinding regions. In LiF, this difference in the delocalization (and associated contraction) characteristics of the two atoms results in a net movement of charge from the lithium towards the fluorine, and accounts for part of the large dipole moment of the molecule. A second factor which influences the direction of delocalization is the difference in sizes of the two atoms. Because the overlap of two atomic wave functions is in general a maximum closer to the smaller atom than to the larger, the constructive interference of the wave functions results in a displacement of charge from the larger atom towards the smaller. In LiF this displacement reinforces the net movement of ($\sigma$) charge from the lithium towards the fluorine. In HF, on the other hand, the two factors act in opposite directions, the relative-size factor causing a displacement of charge towards the smaller, and less electronegative, atom. Fig. 7.5(a) shows that the resulting density-difference distribution between the nuclei is essentially covalent in character, and that the dipole moment (1.8 D) of the molecule is mainly due to the electronic charge on the far side of the fluorine. In LiH the large dipole moment of 6.0 D is due almost entirely to the relative-size factor. Thus, the delocalization effect results in a one-way transfer of charge from both atoms into the region between the nuclei, but the much smaller size of the hydrogen atom causes a large displacement of this charge towards the hydrogen (fig. 7.5(b)).

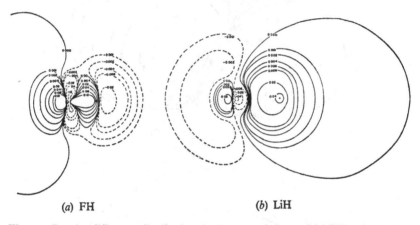

(a) FH                                              (b) LiH

Fig. 7.5. Density-difference distributions in the ground states of (a) HF and (b) LiH (Bader, Keaveny and Cade 1967).

We consider now the contribution of the $\pi$ distribution to the bonding in LiF and HF, and in other '$\sigma$-bonded' molecules. Whereas the fluorine $p\pi$ lone-pair orbitals in $F_2$ are displaced into the antibinding regions and make no effective contribution to the bonding, we have already noted (see table 6.11) that the back-polarization of these orbitals towards the lithium in LiF results in a substantial stabilization of the molecule, and a decrease in the magnitude of the dipole moment (from 7.1 D for basis 3 in table 6.11 to 6.4 D for basis 4). Further evidence for the more general participation of lone pairs is provided by the density-difference distribution of the

## The chemical bond

1$\pi$ molecular orbital of HF shown in fig. 7.6(a) (Kern and Karplus 1964) and by the orbital density distribution of the 1$\pi$ molecular orbital of BF in fig. 7.6(b) (Huo 1965). Both of these show a significant transfer of $\pi$ charge from the nucleus into the region between the nuclei and onto the other atom. This delocalization is accompanied by a decrease of the kinetic energy of the $\pi$ electrons, and allows a further stabilizing contraction towards the fluorine nucleus. The associated change of the dipole moment gives a measure of the extent of the delocalization. For example, the dipole

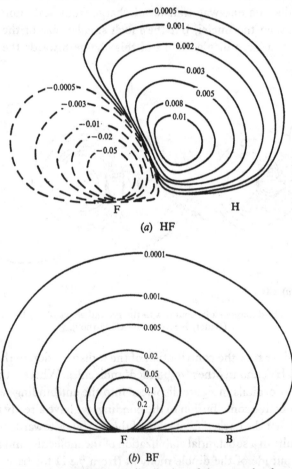

(a) HF

(b) BF

Fig. 7.6. (a) Orbital density-difference distribution of the 1$\pi$ molecular orbital in the ground state of HF (adapted from Kern and Karplus 1964). (b) Orbital density distribution of the 1$\pi$ molecular orbital in the ground state of BF (Huo 1965).

184

moment of HF can be written as

$$\mu(\mathrm{HF}) = \mu(\mathrm{H^+ + F^-}) + \mu_\sigma + \mu_\pi$$

in which $\mu(\mathrm{H^+ + F^-}) = 4.4\,\mathrm{D}$ corresponds to the complete transfer of the electron on H into the $2p\sigma$ atomic orbital of F, $\mu_\sigma = -2.1\,\mathrm{D}$ corrects for the incomplete transfer of this electron, and $\mu_\pi = -0.5\,\mathrm{D}$ is the additional correction due to the delocalization of the $\pi$ distribution (Kern and Karplus 1964). In BF the presence of the valence lone pair on the boron gives a further negative contribution to $\mu_\sigma$ which, at the equilibrium distance of $2.4a_\infty$, is sufficient to reverse the direction of the dipole moment. Fig. 7.7 shows however that the value of the dipole moment is highly sensitive to changes of internuclear distance, and that it has the expected polarity $\mathrm{B^+F^-}$ for distances greater than $2.7a_\infty$, due to the smaller polarization of the boron lone-pair and fluorine $2p\pi$ orbitals. A similar behaviour is observed for the dipole moment of CO (Huo 1965).

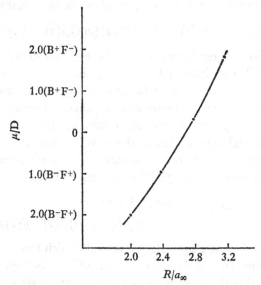

Fig. 7.7. Dependence on the internuclear distance of the dipole moment of BF (Huo 1965).

### 7.2.4 *Electron correlation*

The dissociation energy of a molecule can be written as

$$D_\mathrm{e} = D_\mathrm{e}^{\mathrm{RHF}} + D_\mathrm{e}^{\mathrm{C}}$$

where $D_\mathrm{e}^{\mathrm{RHF}}$ is the dissociation energy in RHF theory and $D_\mathrm{e}^{\mathrm{C}}$ is the contribution of electron correlation. The virial theorem applies to both parts separately and, since $D_e^C$ is always positive, the redistribution of charge due to electron correlation results in a decrease of the potential energy and half as large an increase of the kinetic energy. In a localized-orbital description of the electron distribution, the major contributions to $D_e^C$ come from intrapair correlation in the localized bond orbitals, with lesser contributions from interpair bond–bond and bond–lone pair correlation. Other contributions are normally small.

Electron correlation in a localized two-centre bond can be interpreted in terms of three effects: 'in–out' correlation to describe the tendency of two electrons to be at different distances from the bond axis, angular correlation to describe their tendency to be on opposite sides of the bond axis, and 'left–right' correlation to describe the tendency of the two electrons to be at opposite ends of the bond. We consider the ground state of $H_2$ as the simplest, and most fully investigated, example of the two-electron bond. The RHF wave function for the system

$$\Psi_{RHF}(1,2) = 1\sigma_g(1)\, 1\sigma_g(2) \cdot (1/\sqrt{2})\, [\alpha(1)\,\beta(2) - \beta(1)\,\alpha(2)] \quad (7.13)$$

gives a satisfactory representation of the electron distribution at and near the equilibrium internuclear distance but, as we saw in §3.6, the incorrect behaviour of this wave function at large internuclear distances makes it unsuitable for a description of the process of bond formation and dissociation. The correct behaviour at large distances requires the in-clusion of left–right correlation in the wave function and, as shown in §3.6, this can be achieved by a relaxation of the double-occupancy and symmetry constraints in (7.13), The resulting EHF wave function is

$$\Psi_{EHF}(1,2) = N[\sigma_A(1)\,\sigma_B(2) + \sigma_B(1)\,\sigma_A(2)]$$
$$\times (1/\sqrt{2})\, [\alpha(1)\,\beta(2) - \beta(1)\,\alpha(2)] \quad (7.14)$$

in which $\sigma_A$ and $\sigma_B$ are equivalent $\sigma$ orbitals which transform into each other under inversion of the molecule. At sufficiently large internuclear distances (7.14) is the exact wave function for the system of non-inter-acting ground-state atoms, with spatial part

$$(1/\sqrt{2})\, [1s_A(1)\, 1s_B(2) + 1s_B(1)\, 1s_A(2)]$$

but as the internuclear distance is reduced the orbital on each hydrogen becomes progressively more polarized (delocalized) towards the other. The resulting orbitals at the equilibrium distance are compared in fig. 7.8 with the $1\sigma_g$ orbital of RHF theory (Davidson and Jones 1962). The

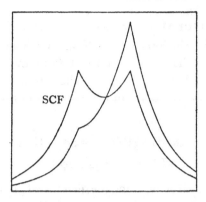

Fig. 7.8. The ground state of $H_2$. SCF and left–right correlated orbitals evaluated along the internuclear axis (Davidson and Jones 1962).

corresponding density-difference distribution

$$\Delta P_{corr}(\boldsymbol{r}) = P_{EHF}(\boldsymbol{r}) - P_{RHF}(\boldsymbol{r})$$

in fig. 7.9 (Bader and Chandra 1968) confirms that the effect of left–right correlation is to correct for the excessive accumulation of charge at the centre of the bond in RHF theory by a partial relocalization nearer the nuclei, thereby reducing the electron-interaction energy. This effect is of particular importance in a molecule like $F_2$, for which the negative value of $D_e^{RHF}$ corresponds to an increase of potential energy and a decrease of kinetic energy on bond formation. The partial relocalization of charge is accompanied by energy changes that are similar to those produced by contraction, and these changes in $F_2$ reverse the signs of the potential-energy and kinetic-energy contributions to the dissociation energy.

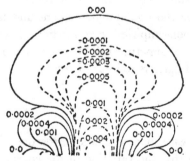

Fig. 7.9. Electron-correlation density distribution $\Delta P_{corr}$ at $R = 1.4a_{\infty}$ in the ground state of $H_2$ (Bader and Chandra 1968; reproduced by permission of the National Research Council of Canada).

# The chemical bond

## 7.3 CONFORMATIONAL CHANGES

A molecule X–Y in which the groups X and Y are linked by a 'single $\sigma$ bond' can undergo changes of conformation by the rotation about the bond of one group relative to the other. The variation of the total energy that accompanies the rotation can be expressed in general as a Fourier expansion,

$$E(\phi) = \tfrac{1}{2}V_1(1 - \cos \phi) + \tfrac{1}{2}V_2(1 - \cos 2\phi) + \tfrac{1}{2}V_3(1 - \cos 3\phi) + \dots$$
$$+ V_1'\sin \phi + V_2'\sin 2\phi + \dots \qquad (7.15)$$

where $\phi$ is the angle of rotation, and the quantities $V_i$ and $V_i'$ determine the relative energies of the stable conformations (local energy minima), the heights of the barriers between them (energy maxima) and the shape of the potential-energy curve. In many molecules a good approximation to $E(\phi)$ is obtained with a very small number of terms. For example, the internal rotation in simple molecules with a three-fold barrier, such as ethane, methylamine and methanol, is accurately described by the potential-energy function

$$E(\phi) = \tfrac{1}{2}V_3(1 - \cos 3\phi) \qquad (7.16)$$

$V_3$ is then the height of the barriers between the equivalent stable conformations of these molecules. The appropriate function for hydrazine, hydroxylamine and hydrogen peroxide is

$$E(\phi) = \tfrac{1}{2}V_1(1 - \cos \phi) + \tfrac{1}{2}V_2(1 - \cos 2\phi) + \tfrac{1}{2}V_3(1 - \cos 3\phi) \qquad (7.17)$$

whilst the remaining two terms shown in (7.15) are required for asymmetric molecules such as the substituted hydrazines and $N$-substituted hydroxylamines (Radom, Hehre and Pople 1972).

We consider here the two systems, ethane and hydrogen peroxide, for which recent non-empirical computations have been sufficiently accurate and complete to permit valid conclusions to be drawn as to the nature and origin of barriers to internal rotation.

## 7.3.1 Ethane

Thermodynamic (heat capacity) and spectroscopic (infrared) studies of ethane have shown that the variation of the energy on rotation of one methyl group relative to the other is given by (7.16), with the staggered conformation ($\phi = 0°$) more stable than the eclipsed conformation

($\phi = 60°$) by about 12 kJ mol$^{-1}$ (Pitzer 1951; Weiss and Leroi 1968). The results of several SCF (RHF) calculations in terms of a variety of basis sets are summarized in table 7.4. It is clear from these that RHF theory is adequate for the prediction of the barrier height, in agreement with Freed's theorem (Freed 1968; Allen and Arents 1972) that conformational barriers in Hartree–Fock theory are accurate to second order if the potential-energy curve is sinusoidal, as is the case for ethane (Freed's theorem is a consequence of Brillouin's theorem, and the same conditions apply). A second conclusion to be drawn from table 7.4 is that the computed barrier is rather insensitive to the basis used to construct the molecular orbitals. For example, the barriers obtained with a minimal Slater-function basis (calculations 2 and 3 in table 7.4) and with a large contracted-Gaussian basis (calculations 7) differ by only 0.3 kJ mol$^{-1}$, whilst the total energies differ by about 700 kJ mol$^{-1}$. This basis-independence appears to be more generally true for molecules with no valence lone pairs on the two atoms linked by the bond about which the rotation occurs.

Whereas the results in table 7.4 (particularly sets 6 and 7) show that allowing the nuclear structure to relax during the rotation gives only a small lowering of the computed barrier, the decomposition of the total energy into kinetic-energy and potential-energy components (table 7.5) suggests that this relaxation is accompanied by a substantial reorganization of the electron distribution. The difference between the total energies of the eclipsed and staggered conformations can be written as

$$\Delta E = E_{\text{eclipsed}} - E_{\text{staggered}}$$
$$= \Delta T + \Delta V_{nn} + \Delta V_{ne} + \Delta V_{ee}$$

where $\Delta T$ is the change in the kinetic energy of the electrons on going from the staggered to the eclipsed conformation, $\Delta V_{nn}$ is the change in the nuclear-repulsion energy, $\Delta V_{ne}$ is the change in the nucleus–electron attraction energy, and $\Delta V_{ee}$ is the change in the electron–repulsion energy. According to the virial theorem (§7.1),

$$\Delta T = -\Delta E$$
$$\Delta V = \Delta V_{nn} + \Delta V_{ne} + \Delta V_{ee} = 2\Delta E$$

so that the increase in total energy on going from the staggered to the eclipsed conformation is made up of a decrease in the kinetic energy of the electrons and an increase in the (total) potential energy.

The virial theorem is approximately satisfied for the non-rigid rotation

TABLE 7.4 *Hindered rotation in ethane. Total SCF energies of the staggered and eclipsed conformations, barrier height ($|\Delta E|$), and changes in the C–C and C–H bond distances and in the HCH bond angle (these geometric parameters are unchanged in a rigid rotation)*

| | $E/H_\infty$ | | Barrier | Changes in geometry (staggered to eclipsed) | | |
|---|---|---|---|---|---|---|
| | Staggered | Eclipsed | $|\Delta E|$/kJ mol$^{-1}$ | $\Delta R_{CC}/a_\infty$ | $\Delta R_{CH}/a_\infty$ | $\Delta\angle$HCH |
| 1 Blustin and Linnett 1974 | −67.0276 | −67.0227 | 12.9 | +0.049 | −0.002 | −0.4° |
| 2 Pitzer and Lipscomb 1963 | −78.9912 | −78.9859 | 13.7 | | Rigid rotation | |
| 3 Stevens 1970 | −79.0999 | −79.0946 | 13.7 | +0.038 | −0.006 | −0.3° |
| 4 Fink and Allen 1967 | −79.1478 | −79.1438 | 10.5 | | Rigid rotation | |
| 5 Clementi and von Niessen 1971 | −79.2031 | −79.1981 | 13.1 | | Rigid rotation | |
| 6a Veillard 1970 | −79.2377 | −79.2319 | 15.2 | | Rigid rotation | |
| 6b Veillard 1970 | −79.2390 | −79.2341 | 12.9 | +0.036 | 0 | −0.3° |
| 7a Clementi and Popkie 1972 | −79.2581 | −79.2529 | 14.1 | | Rigid rotation | |
| 7b Clementi and Popkie 1972 | −79.2588 | −79.2536 | 13.4 | +0.035 | 0 | −0.3° |
| Experiment (Weiss and Leroi 1968) | | | 12.25 ± 0.11 | | | |

TABLE 7.5   *Effect of nuclear relaxation on the barrier to internal rotation in ethane. Energies are in units of* kJ mol$^{-1}$

| Energy changes (staggered to eclipsed) | Rigid rotation† | Non-rigid rotation‡ |
|---|---|---|
| Nuclear repulsion, $\Delta V_{nn}$ | +19.7 | −699.1 |
| Electron repulsion, $\Delta V_{ee}$ | +69.8 | −644.8 |
| Nucleus-electron attraction, $\Delta V_{ne}$ | −128.6 | +1374.0 |
| Total potential, $\Delta V$ | −39.2 | +30.1 |
| Kinetic, $\Delta T$ | +52.9 | −17.2 |
| Total, $\Delta E = \Delta T + \Delta V$ | +13.7 | +12.9 |

† Minimal STO (Pitzer and Lipscomb 1963; Epstein and Lipscomb 1970).
‡ Extended FSGO (Blustin and Linnett 1974).

in table 7.5, but cannot in general be satisfied for a rigid rotation. Table 7.5 shows that the kinetic-energy and potential-energy changes that accompany a rigid rotation have *opposite* sign from those that accompany a non-rigid rotation.† Thus, a rigid rotation from the staggered to the eclipsed conformation is accompanied by the 'expected', but incorrect, increases in nuclear–repulsion and electron–repulsion energies ($\Delta V_{nn} > 0$, $\Delta V_{ee} > 0$), as well as by an increase in the kinetic energy ($\Delta T > 0$), and a large compensating lowering ($\Delta V_{ne} < 0$) of the nucleus–electron attraction energy. The relaxation of the nuclear structure, however, reverses the signs of these energy changes. A *non-rigid* rotation from the staggered to the eclipsed conformation is accompanied by *decreases* in the nuclear–repulsion and electron–repulsion energies ($\Delta V_{nn} < 0$, $\Delta V_{ee} < 0$), a decrease in the kinetic energy ($\Delta T < 0$, as required by the virial theorem), and a large destabilizing increase ($\Delta V_{ne} > 0$) of the nucleus–electron interaction energy.

The energy changes for both rigid and non-rigid rotation can be interpreted in terms of the delocalization and contraction effects discussed in §7.2. Thus, the energy changes for a rigid rotation from the *eclipsed* to the *staggered* conformation can be ascribed to

(i) A decrease in the exchange, or Pauli-principle, repulsion (§7.2.2) between the C–H bond electron pairs on different methyl groups, and

(ii) A delocalization effect involving a movement of electronic charge parallel to the direction of the C–C axis from the regions around the C–H

† Although the energy changes in table 7.5 have been obtained from very approximate SCF calculations, more accurate work has shown that they have the correct signs and orders of magnitude.

bonds of the methyl groups into the region between the methyl groups; that is, an increase in 'hyperconjugation'.

The decrease in the exchange repulsion between the methyl groups on going from the eclipsed to the staggered conformation can be regarded as the 'origin' of the barrier; that is, as the cause that makes the delocalization effect possible. As in the case of $H_2^+$ (§7.2.1), the delocalization effect leads to a decrease in the kinetic energy of the electrons and a (smaller) increase in the potential energy. The additional changes that occur when the nuclear structure is allowed to relax can be ascribed to an overall contraction of the electron distribution with the nuclei, particularly the carbons, as contraction centres. This contraction results in increases of the kinetic energy and electron–repulsion energies, and a large decrease ($\Delta V_{ne} < 0$) of the nucleus–electron attraction energy. In addition, the changes in bond lengths and bond angles (table 7.4) suggest that the contraction is similar to that in a diatomic molecule like $F_2$, with a strong contraction in the C–C bond region and a weaker contraction, or even expansion, in the regions behind the carbons (inside the methyl umbrellas).

### 7.3.2 *Hydrogen peroxide*

Infrared studies of the internal rotation in hydrogen peroxide have shown that the dependence of the energy on the dihedral (HOOH) angle $\phi$ can be described to good approximation by the function (7.17),

$$E(\phi) = E_1(\phi) + E_2(\phi) + E_3(\phi)$$

where

$$E_n(\phi) = \tfrac{1}{2}V_n(1 - \cos n\phi)$$

with $V_1 = -25.9\,\text{kJ mol}^{-1}$, $V_2 = -16.0\,\text{kJ mol}^{-1}$ and $V_3 = -1.1\,\text{kJ mol}^{-1}$ (Hunt *et al.* 1965; Ewig and Harris 1970). The total-energy curve and its Fourier components are plotted in fig. 7.10. The total energy has maxima for the eclipsed (*cis*) and staggered (*trans*) conformations, and a minimum for the dihedral angle $\phi = 111.5°$. The *cis* and *trans* barrier heights are $31.6\,\text{kJ mol}^{-1}$ and $4.6\,\text{kJ mol}^{-1}$ respectively. Fig. 7.10 shows that the three-fold component $E_3(\phi)$ makes only a very small contribution to the total energy, and that the stable geometry and the barriers are determined by a balance between the one-fold component $E_1(\phi)$, which favours the staggered conformation, and the two-fold component $E_2(\phi)$, which favours the angle $\phi = 90°$ with the two hydroxyl groups perpendicular to each other.

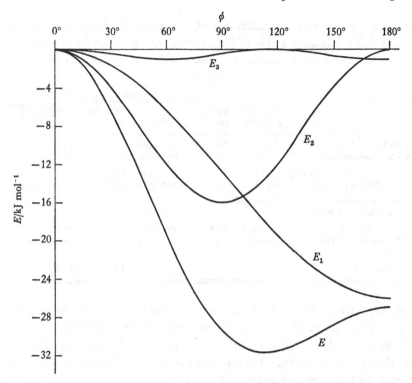

Fig. 7.10. Internal rotation in $H_2O_2$. Dependence on the dihedral angle $\phi$ of the total energy $E$ and of its Fourier components $E_n$.

The results of several SCF calculations in terms of a variety of basis sets are summarized in table 7.6. Unlike the case of ethane, a satisfactory description of the internal rotation in hydrogen peroxide is obtained only when polarization functions are included in the basis, with no *trans* barrier predicted in the absence of these functions. This result is typical of molecules which have valence lone pairs on the atoms linked by the bond about which the rotation occurs. It is also then true of other conformational changes, such as the inversion of ammonia, for which the pyramidal conformation is found to be more stable than the planar conformation only when polarization functions are included in the basis (Rauk, Allen and Clementi 1970). As in the case of ethane, on the other hand, allowing the nuclear structure to relax during the internal rotation in hydrogen peroxide results in only small, though significant, changes in the barrier heights, but is accompanied by large changes in the kinetic-energy and potential-energy components, with the same trends as shown in table 7.5

TABLE 7.6 *Hindered rotation in hydrogen peroxide. Total SCF energy and dihedral angle of the stable conformation, and barrier heights.*

| | Type of rotation | Polarization functions in basis set | Stable conformation | | Barriers ($\Delta E$/kJ mol$^{-1}$) | |
|---|---|---|---|---|---|---|
| | | | $-E/H_\infty$ | Dihedral angle | cis | trans |
| 1 Palke and Pitzer 1967 | Rigid | No | 150.1467 | 180° | 39.6 | |
| 2 Davidson and Allen 1971 | Rigid | No | 150.7393 | 180° | 77.0 | |
| 3 Davidson and Allen 1971 | Rigid | Yes | 150.7910 | 132° | 58.0 | 1.0 |
| 4 Guidotti *et al.* 1972 | Rigid | Yes | 150.8319 | 120° | 45.6 | 3.0 |
| 5 Stevens 1970 | Non-rigid | No | 150.2353 | 180° | 39.9 | |
| 6 Davidson and Allen 1971 | Non-rigid | No | 150.7049 | 180° | 74.1 | |
| 7 Veillard 1970 | Non-rigid | Yes | 150.7993 | 123° | 45.6 | 2.5 |
| 8 Dunning and Winter 1971 | Non-rigid | Yes | 150.8219 | 113.7° | 34.9 | 4.6 |
| Experiment | | | | 111.5° | 31.6 | 4.6 |

(Veillard 1970). On going from the *cis* to the stable conformation, the O–O bond length decreases by about $0.04a_\infty$ and the OOH bond angle decreases by about 5°. These geometric parameters then remain relatively unchanged on going from the stable to the *trans* conformation, reflecting the small (total) energy changes in this region.

The interpretation of the internal rotation in hydrogen peroxide follows the same lines as that in ethane, but with obvious complications due to the presence of the lone pairs. The variation of the two-fold component $E_2(\phi)$ of the energy (fig. 7.10) is most readily interpreted in terms of a localized-orbital description of the electron distribution in which one valence lone pair of electrons on each oxygen occupies a $2p$ atomic orbital at right angles to the OOH plane, whilst the second lone-pair orbital lies in the OOH plane and has a large amount of $s$ character. The exchange repulsion between the two pairs of electrons in the $2p$ orbitals is then greatest for the planar (*cis* and *trans*) conformations, and a minimum for the perpendicular conformation. The exchange repulsion between the second pair of lone-pair electrons, on the other hand, is a minimum for the planar *trans* conformation and a maximum for the *cis* conformation. The bond–bond exchange repulsion also favours the *trans* conformation, whilst the bond–lone pair repulsion favours the *cis* conformation. As shown in fig. 7.10, the resulting one-fold component $E_1(\phi)$ has its minimum for the *trans* conformation.

# References

Aarons L. J., Guest M. F., Hall M. B. and Hillier I. H. (1973). *J.C.S. Faraday* II **69**, 563

Abramowitz M. and Stegun I. A. (1965). *Handbook of Mathematical Functions*. Dover, New York

Ahlrichs R. (1973). *Chem. Phys. Letters* **19**, 174

Allen L. C. and Arents J. (1972). *J. Chem. Phys.* **57**, 1818

Aung S., Pitzer R. M. and Chan S. I. (1968). *J. Chem. Phys.* **49**, 2071

Bader R. F. W. and Bandrauk A. D. (1968). *J. Chem. Phys.* **49**, 1653

Bader R. F. W. and Chandra A. K. (1968). *Can. J. Chem.* **46**, 953

Bader R. F. W. and Henneker W. H. (1965). *J. Amer. Chem. Soc.* **87**, 3063

Bader R. F. W., Henneker W. H. and Cade P. E. (1967). *J. Chem. Phys.* **46**, 3341

Bader R. F. W., Keaveny I. and Cade P. E. (1967). *J. Chem. Phys.* **47**, 3381

Bagus P. S. and Schaefer H. F. (1972). *J. Chem. Phys.* **56**, 224

Basch H. (1970). *Chem. Phys. Letters* **5**, 337

Bender C. F. and Schaefer H. F. (1970). *J. Amer. Chem. Soc.* **92**, 4984

Bender C. F., Schaefer H. F., Franceschetti D. R. and Allen L. C. (1972). *J. Amer. Chem. Soc.* **94**, 6888

Berlin T. (1951). *J. Chem. Phys.* **19**, 208

Bernheim R. A., Bernard H. W., Wang P. S., Wood L. S. and Skell P. S. (1970). *J. Chem. Phys.* **53**, 1280

Bethe H. A. and Salpeter E. E. (1957). *Quantum Mechanics of One- and Two-Electron Atoms*. Springer-Verlag, Berlin

Bethell D. (1973). *Organic Reactive Intermediates*. Ed. S. P. McManus. Academic Press, New York and London

Bishop D. M. (1973). *Group Theory and Chemistry*. Clarendon Press, Oxford

Blustin P. H. and Linnett J. W. (1974). *J.C.S. Faraday* II **70**, 290

Boer F. P. and Lipscomb W. N. (1969). *J. Chem. Phys.* **50**, 989.

Boys S. F. (1950). *Proc. Roy. Soc.* A **200**, 542

Boys S. F. (1960). *Rev. Mod. Phys.* **32**, 296

Brillouin L. (1933). *Actualites Sci. Ind.* **71**; (1934) **159**

Brundle C. R., Robin M. B. and Basch H. (1970). *J. Chem. Phys.* **53**, 2196

Bunge C. F. (1967). *Phys. Rev.* **154**, 70

Cade P. E., Bader R. F. W., Henneker W. H. and Keaveny I. (1969). *J. Chem. Phys.* **50**, 5313

Cade P. E. and Huo W. M. (1967a). *J. Chem. Phys.* **47**, 614

Cade P. E. and Huo W. M. (1967b). *J. Chem. Phys.* **47**, 649

Cade P. E., Sales K. D. and Wahl A. C. (1966). *J. Chem. Phys.* **44**, 1973

Carr R. W., Eder T. W. and Topor M. G. (1970). *J. Chem. Phys.* **53**, 4716

Christoffersen R. E. (1971). *J. Amer. Chem. Soc.* **93**, 4104

Christoffersen R. E. (1972). *Adv. Quantum Chem.* **6**, 333

Christoffersen R. E. (1973). *Intern. J. Quantum Chem.* 7S, 169

Christoffersen R. E., Genson D. W. and Maggiora G. M. (1971). *J. Chem. Phys.* **54**, 239

Clark D. T. and Scanlan I. W. (1974). *J.C.S. Faraday* II **70**, 1222

Clementi E. (1962). *J. Chem. Phys.* **36**, 33

Clementi E. (1965a). *IBM J. Res. Develop.* **9**, 2

Clementi E. (1965b). *Tables of Atomic Functions*. Special IBM Technical Report, IBM, San Jose

Clementi E. (1967a). *J. Chem. Phys.* **46**, 4731

Clementi E. (1967b). *J. Chem. Phys.* **47**, 4485

# References

Clementi E. (1968). *Chem. Rev.* **68**, 341
Clementi E., Andre J. M., Andre M. C., Klint D. and Hahn D. (1969). *Acta Phys. Acad. Sci. Hung.* **27**, 493
Clementi E., Clementi H. and Davis D. R. (1967). *J. Chem. Phys.* **46**, 4725
Clementi E. and Popkie H. (1972). *J. Chem. Phys.* **57**, 4870
Clementi E. and Raimondi D. L. (1963). *J. Chem. Phys.* **38**, 2686
Clementi E. and von Niessen W. (1971). *J. Chem. Phys.* **54**, 521
Cotton F. A. (1963). *Chemical Applications of Group Theory.* Interscience, New York
Coulson C. A. (1961). *Valence.* Oxford University Press, London
Das G. and Wahl A. C. (1966). *J. Chem. Phys.* **44**, 87
Das G. and Wahl A. C. (1972a). *J. Chem. Phys.* **56**, 1769
Das G. and Wahl A. C. (1972b). *J. Chem. Phys.* **56**, 3532
Davidson E. R. (1972). *Rev. Mod. Phys.* **44**, 451
Davidson E. R. and Jones L. L. (1962). *J. Chem. Phys.* **37**, 1918
Davidson R. B. and Allen L. C. (1971). *J. Chem. Phys.* **55**, 519
Davis D. W., Hollander J. M., Shirley D. A. and Thomas T. D. (1970). *J. Chem. Phys.* **52**, 3295
Davis D. W., Shirley D. A. and Thomas T. D. (1972). *J. Amer. Chem. Soc.* **94**, 6565
Dewar M. J. S. (1969). *The Molecular Orbital Theory of Organic Chemistry.* McGraw-Hill, New York
Ditchfield R., Hehre W. J. and Pople J. A. (1971). *J. Chem. Phys.* **54**, 724
Dunning T. H. (1970a). *J. Chem. Phys.* **53**, 2823
Dunning T. H. (1970b). *Chem. Phys. Letters* **7**, 423
Dunning T. H. (1971a). *J. Chem. Phys.* **55**, 3958
Dunning T. H. (1971b). *J. Chem. Phys.* **55**, 716
Dunning T. H., Pitzer R. M. and Aung S. (1972). *J. Chem. Phys.* **57**, 5044
Dunning T. H. and Winter N. W. (1971). *Chem. Phys. Letters* **11**, 194
Edmiston C. and Ruedenberg K. (1963). *Rev. Mod. Phys.* **35**, 457
Edmiston C. and Ruedenberg K. (1966). *Quantum Theory of Atoms, Molecules, and the Solid State.* Ed. P. O. Löwdin. Academic Press, New York, p. 263
England W. and Ruedenberg K. (1971). *Theoret. Chim. Acta* **22**, 196
Epstein I. R. and Lipscomb W. N. (1970). *J. Amer. Chem. Soc.* **92**, 6094
Epstein I. R., Marynick D. S. and Lipscomb W. N. (1973). *J. Amer. Chem. Soc.* **95**, 1760
Erickson W. D. and Linnett J. W. (1972). *J.C.S. Faraday* II **68**, 693
Ewig C. S. and Harris D. O. (1970). *J. Chem. Phys.* **52**, 6268
Eyring H., Walter J. and Kimball G. E. (1944). *Quantum Chemistry.* Wiley, New York
Feinberg M. J. and Ruedenberg K. (1971). *J. Chem. Phys.* **54** 1495
Feinberg M. J., Ruedenberg K. and Mehler E. L. (1970). *Adv. Quantum Chem.* **5**, 28
Fink W. H. and Allen L. C. (1967). *J. Chem. Phys.* **46**, 2261
Finkelstein B. N. and Horowitz G. E. (1928). *Z. Physik* **48**, 118
Foster J. M. and Boys S. F. (1960). *Rev. Mod. Phys.* **32**, 305
Freed K. F. (1968). *Chem. Phys. Letters* **2**, 255
Froese C. (1966). *J. Chem. Phys.* **45**, 1417
Frost A. A. (1967). *J. Chem. Phys.* **47**, 3707
Frost A. A. (1968). *J. Phys. Chem.* **72**, 1289
Frost A. A. (1970). *Theoret. Chim. Acta* **18**, 156
Frost A. A. and Rouse R. A. (1968). *J. Amer. Chem. Soc.* **90**, 1965
Gelius U., Roos B. and Siegbahn P. (1970). *Chem. Phys. Letters* **4**, 471
Green S. (1971). *J. Chem. Phys.* **54**, 827
Grimaldi F., Lecourt A. and Moser C. (1967). *Intern. J. Quantum Chem.* **1S**, 153
Guest M. F. and Hillier I. H. (1973). *Mol. Phys.* **26**, 435
Guest M. F. and Hillier I. H. (1974). *J.C.S. Faraday* II **70**, 2004
Guidotti C., Lamanna U., Maestro M. and Moccia R. (1972). *Theoret. Chim. Acta* **27**, 55
Guillemin V. and Zener C. (1929). *Proc. Natl. Acad. Sci. U.S.* **15**, 314

Halberstadt M. L. and McNesby J. R. (1967). *J. Amer. Chem. Soc.* **89**, 3417
Hall G. G. and Lennard-Jones J. E. (1951). *Proc. Roy. Soc.* A **205**, 357
Hall J. H., Marynick D. S. and Lipscomb W. N. (1974). *J. Amer. Chem. Soc.* **96**, 770
Hariharan P. C. and Pople J. A. (1973). *Theoret. Chim. Acta* **28**, 213
Hartmann H. and Clementi E. (1964). *Phys. Rev.* **133**, A1295
Hartree D. R. (1957). *The Calculation of Atomic Structures.* Wiley, New York
Hehre W. J., Ditchfield R. and Pople J. A. (1972). *J. Chem. Phys.* **56**, 2257
Hehre W. J. and Pople J. A. (1970). *J. Amer. Chem. Soc.* **92**, 2191
Hehre W. J., Stewart R. F. and Pople J. A. (1969). *J. Chem. Phys.* **51**, 2657
Herzberg G. (1967). *Electronic Spectra of Polyatomic Molecules.* Van Nostrand, Princeton
Herzberg G. and Johns J. W. C. (1966). *Proc. Roy. Soc.* A **295**, 107
Herzberg G. and Johns J. W. C. (1971). *J. Chem. Phys.* **54**, 2276
Hoffmann R. (1963). *J. Chem. Phys.* **39**, 1397
Hunt R. H., Leacock R. A., Peters C. W. and Hecht K. T. (1965). *J. Chem. Phys.* **42**, 1931
Huo W. (1965). *J. Chem. Phys.* **43**, 624
Huzinaga S. (1965). *J. Chem. Phys.* **42**, 1293
Huzinaga S. (1971). *Approximate Atomic Functions.* Technical Report of the Division of Theoretical Chemistry, University of Alberta
Huzinaga S. and Arnau C. (1970a). *J. Chem. Phys.* **53**, 451
Huzinaga S. and Arnau C. (1970b). *J. Chem. Phys.* **52**, 2224; **53**, 348
Hylleraas E. A. (1929). *Z. Physik* **54**, 347
Hylleraas E. A. (1964). *Adv. Quantum Chem.* **1**, 1
Kern C. W. and Karplus M. (1964). *J. Chem. Phys.* **40**, 1374
Kochanski E. and Lehn J. M. (1969). *Theoret. Chim. Acta* **14**, 281
Kolos W. and Roothaan C. C. J. (1960). *Rev. Mod. Phys.* **32**, 219
Kolos W. and Wolniewicz L. (1964). *J. Chem. Phys.* **41**, 3674
Koopmans T. (1933). *Physica* **1**, 104
Lathan W. A., Curtiss L. A., Hehre W. J., Lisle J. B. and Pople J. A. (1974). *Prog. Phys. Org. Chem.* **11**, 175
Laws E. A., Stevens R. M. and Lipscomb W. N. (1972). *J. Amer. Chem. Soc.* **94**, 4461
Löwdin P. O. (1956). *Adv. Phys.* **5**, 1
Löwdin P. O. (1959a). *Adv. Chem. Phys.* **2**, 207
Löwdin P. O. (1959b). *J. Mol. Spect.* **3**, 46
McLaughlin D. R., Bender C. F. and Schaefer H. F. (1972). *Theoret. Chim. Acta* **25**, 352
McWeeny R. (1972). *Quantum Mechanics. Principles and Formalism.* Pergamon, Oxford.
McWeeny R. (1973). *Quantum Mechanics. Methods and Basic Applications.* Pergamon, Oxford
McWeeny R. and Del Re G. (1968). *Theoret. Chim. Acta* **10**, 13
McWeeny R. and Steiner E. (1965). *Adv. Quantum Chem.* **2**, 93
McWeeny R. & Sutcliffe B. T. (1969). *Methods of Molecular Quantum Mechanics.* Academic Press, London and New York
Magnasco V. and Perico A. (1967). *J. Chem. Phys.* **47**, 971
Maly J. and Hussonois M. (1973). *Theoret. Chim. Acta* **28**, 363
Mann J. B. and Johnson W. R. (1971). *Phys. Rev.* **A4**, 41
Marynick D. S. and Lipscomb W. N. (1972). *J. Amer. Chem. Soc.* **94**, 8692
Matcha R. L. (1968). *J. Chem. Phys.* **48**, 335
Matsen F. A. (1964). *Adv. Quantum Chem.* **1**, 59
Meyer W. (1973). *J. Chem. Phys.* **58**, 1017
Moskowitz J. W., Hollister C., Hornback C. J. and Basch H. (1970). *J. Chem. Phys.* **53**, 2570
Moss R. E. (1973). *Advanced Molecular Quantum Mechanics.* Chapman and Hall, London
Mulliken R. S. (1955). *J. Chem. Phys.* **23**, 1833

# References

Mulliken R. S. (1962). *J. Chem. Phys.* **36**, 3428

Murrell J. N., Kettle S. F. A. and Tedder J. M. (1970). *Valence Theory*. Wiley, London

Naleway C. A. and Schwartz M. E. (1973). *Theoret. Chim. Acta* **30**, 347.

Nesbet R. K. (1962). *J. Chem. Phys.* **36**, 1518

Nesbet R. K. (1967). *Phys. Rev.* **155**, 51

Nesbet R. K. (1969). *Adv. Chem. Phys.* **14**, 1

Newton M. D., Lathan W. A., Hehre W. J. and Pople J. A. (1970*a*). *J. Chem. Phys.* **52**, 4046

Newton M. D. and Switkes E. (1971). *J. Chem. Phys.* **54**, 3179

Newton M. D., Switkes E. and Lipscomb W. N. (1970*b*). *J. Chem. Phys.* **53**, 2645

O'Neil S. V., Schaefer H. F. and Bender C. F. (1971). *J. Chem. Phys.* **55**, 162

Palke W. E. and Pitzer R. M. (1967). *J. Chem. Phys.* **46**, 3948

Pauling L. (1928). *Chem. Rev.* **5**, 173

Pauling L. (1960). *The Nature of the Chemical Bond*. Cornell University Press

Petke J. D., Whitten J. L. and Douglas A. W. (1969). *J. Chem. Phys.* **51**, 256

Pitzer K. S. (1951). *Discuss. Faraday Soc.* **10**, 66

Pitzer R. M. and Lipscomb W. N. (1963). *J. Chem. Phys.* **39**, 1995

Pitzer R. M. and Merrifield D. P. (1970). *J. Chem. Phys.* **52**, 4782

Politzer P. and Harris R. R. (1970). *J. Amer. Chem. Soc.* **92**, 6451

Politzer P. and Mulliken R. S. (1971). *J. Chem. Phys.* **55**, 5135

Politzer P. and Politzer A. (1973). *J. Amer. Chem. Soc.* **95**, 5450

Politzer P. and Reggio P. H. (1972). *J. Amer. Chem. Soc.* **94**, 8308

Pople J. A. (1957). *Quart. Rev.* **11**, 273

Pople J. A. and Beveridge D. L. (1970). *Approximate Molecular Orbital Theory*. McGraw-Hill, New York

Preston H. J. T. and Kaufman J. J. (1973). *Intern. J. Quantum Chem.* **S7**, 207

Radom L., Hehre W. J. and Pople J. A. (1972). *J. Amer. Chem. Soc.* **94**, 2371

Ransil B. J. (1960). *Rev. Mod. Phys.* **32**, 245

Rauk A., Allen L. C. and Clementi E. (1970). *J. Chem. Phys.* **52**, 4133

Roos B. and Siegbahn P. (1970). *Theoret. Chim. Acta* **17**, 209

Roothaan C. C. J. (1951). *Rev. Mod. Phys.* **23**, 69

Roothaan C. C. J. (1960). *Rev. Mod. Phys.* **32**, 179

Rothenberg S. (1969). *J. Chem. Phys.* **51**, 3389

Rothenberg S., Young R. H. and Schaefer H. F. (1970). *J. Amer. Chem. Soc.* **92**, 3243

Ruedenberg K. (1962). *Rev. Mod. Phys.* **34**, 326

Salez C. and Veillard A. (1968). *Theoret. Chim. Acta* **11**, 441

Schaefer H. F. (1972). *The Electronic Structure of Atoms and Molecules*. Addison-Wesley, Reading, Massachusetts

Schwartz M. E. (1970). *Chem. Phys. Letters* **6**, 631

Schwartz M. E. and Allen L. C. (1970). *J. Amer. Chem. Soc.* **92**, 1466

Schwenzer G. M., Liskow D. H., Schaefer H. F., Bagus P. S., Liu B., McLean A. D. and Yoshimine M. (1973). *J. Chem. Phys.* **58**, 3181

Shavitt I. (1963). *Methods in Computational Physics* **2**, 1. Academic Press, New York and London

Shih S., Buenker R. J., Peyerimhoff S. D. and Wirsam B. (1970). *Theoret. Chim. Acta* **18**, 277

Shipman L. L. and Christoffersen R. E. (1973). *J. Amer. Chem. Soc.* **95**, 4733

Shirley D. A. (1973). *Adv. Chem. Phys.* **23**, 85

Shull H. and Hall G. G. (1959). *Nature, Lond.* **184**, 1559

Siegbahn K., Nordling C., Johansson G., Hedman J., Heden P. F., Hamrin K., Gelius U., Bergmark T., Werme L. O., Manne R. and Baer Y. (1969). *ESCA Applied to Free Molecules*. North-Holland, Amsterdam

Sinanoglu O. (1964). *Adv. Chem. Phys.* **6**, 315

Sinanoglu O. (1969). *Adv. Chem. Phys.* **14**, 237

# References

Slater J. C. (1963). *Quantum Theory of Molecules and Solids* vol. 1. McGraw-Hill, New York

Snyder L. C. and Basch H. (1969). *J. Amer. Chem. Soc.* **91**, 2189

Snyder L. C. and Basch H. (1972). *Molecular Wave Functions and Properties*. Wiley, New York

Staemmler V. (1973). *Theoret. Chim. Acta* **31**, 49

Steiner E. and Sykes S. (1972). *Mol. Phys.* **23**, 643

Stevens R. M. (1970). *J. Chem. Phys.* **52**, 1397

Stevens R. M., Switkes E., Laws E. A. and Lipscomb W. N. (1971). *J. Amer. Chem. Soc.* **93**, 2603

Stewart R. F. (1970). *J. Chem. Phys.* **52**, 431

Switkes E., Epstein I. R., Tossell J. A., Stevens R. M. and Lipscomb W. N. (1970*a*) *J. Amer. Chem. Soc.* **92**, 3837

Switkes E., Lipscomb W. N. and Newton M. D. (1970*b*). *J. Amer. Chem. Soc.* **92**, 3847

Switkes E., Stevens R. M., Lipscomb W. N. and Newton M. D. (1969). *J. Chem. Phys.* **51**, 2085

Taylor G. R. and Parr R. G. (1952). *Proc. Natl. Acad. Sci. U.S.* **38**, 154

Turner D. W. (1968). *Chemistry in Britain* **4**, 435

Veillard A. (1968). *Theoret. Chim. Acta* **12**, 405

Veillard A. (1970). *Theoret. Chim. Acta* **18**, 21

Veillard A. and Clementi E. (1967). *Theoret. Chim. Acta* **7**, 133

Veillard A. and Clementi E. (1968). *J. Chem. Phys.* **49**, 2415

von Niessen W. (1972). *J. Chem. Phys.* **56**, 4290

Wachters A. J. H. (1970). *J. Chem. Phys.* **52**, 1033

Wahl A. C. (1964). *J. Chem. Phys.* **41**, 2600

Wasserman E., Yager W. A. and Kuck V. (1970). *Chem. Phys. Letters* **7**, 409

Weinbaum S. (1933). *J. Chem. Phys.* **1**, 593

Weiss S. and Leroi G. E. (1968). *J. Chem. Phys.* **48**, 962

Whitman D. R. and Hornback C. J. (1969). *J. Chem. Phys.* **51**, 398

Whitten J. L. (1963). *J. Chem. Phys.* **39**, 349

Whitten J. L. (1966). *J. Chem. Phys.* **44**, 359

Wigner E. P. (1959). *Group Theory and its Applications to the Quantum Mechanics of Atomic Spectra*. Academic Press, New York and London

Williams J. E. and Streitwieser A. (1974). *Chem. Phys. Letters* **25**, 507

Wilson C. W. and Goddard W. A. (1972). *Theoret. Chim. Acta* **26**, 195

Yoshimine M. and McLean A. D. (1967). *Intern. J. Quantum. Chem.* **1S**, 313

# Index

# Index

determinant
  secular, 6, 81
  Slater, 38
diborane, 138, 139
dipole moment, 56–7, 118, 119
dissociation energy
  and electron correlation, 185–7
  in Hartree–Fock, theory, 58–9
double-occupancy constraint, 46–7, 48
  relaxation of, 58, 61, 63
double-zeta basis (DZ), 84, 98–101

Edmiston–Ruedenberg localization
  method, 131–3
effective atomic charges, 151
eigenfunctions
  complete set of, 3, 11
  of Hamiltonian, 3
  of orbital angular momentum, 31–4
  of spin, 34–6
electron binding energies
  inner-shell, 159–68
  and Koopmans' theorem, 53–4
electron correlation, 62–76, 185–7
  perturbative treatment of, 67–73
  see also correlation
electron permutation symmetry, 13,
    24–7
equivalent orbitals, 129
equivalent representations, 19
ethane, internal rotation in, 188–92
ethylene, 137–8, 144
exchange integral, 41
exchange operator, 45, 46–7
exchange repulsion, 179
excited configuration, 50, 55–6
excited states, 50, 101–6
expansion
  in complete set, 3, 8–9, 24
  in Slater determinants, 51–2, 55
  see also configuration interaction
expectation value
  definition of, 117
  of dipole moment, 56
  of Hamiltonian, 41, 118
  of one-electron operator, 119
extended Hartree–Fock theory (EHF),
    58, 62–4

Fermi contact term, 58
Fermi–Dirac statistics, 27
Fermi hole, 64
fermion, 27
Finkelstein–Horowitz wave function,
    172
floating Gaussians, 107

floating spherical Gaussian orbital
  method (FSGO), 107–10
fluorine molecule
  dissociation energy of, 59, 60, 75
  electron correlation in, 75–6, 187

Gaussian lobe functions, 90
Gaussian representation of Slater-type
  orbitals (STO–$N$G), 96–7
Gaussians
  contracted, 92–5
  floating, 107
  primitive, 94
Gaussian-type orbitals (GTO), 88–95
geometry, see molecular geometry
gross atomic population, 150
group theory, 13–24
  and the Schrödinger equation, 22–3
Guillemin–Zener wave function, 172

Hamiltonian
  atomic, 2
  in Born–Oppenheimer approximation,
    2–4
  Hartree–Fock, 44, 47, 49, 52
  non-empirical, 1
  one-electron, 37
Hartree energy, 3–5
Hartree–Fock equations
  restricted for closed shells, 49, 133
  restricted for open shells, 52
  unconstrained, 44, 47
Hartree–Fock theory (model), 40–61,
    80–3
  constraints in, 45–8
  extended (EHF), 58, 62–4
  matrix representation of, 80–3
  projected (PHF), 58
  restricted (RHF), 48–53
  spin-unrestricted (SUHF), 58
  unconstrained, 44–5
helium atom
  electron correlation in, 62–5
  orbital model of, 43–4
helium molecule, 178–9
Hellmann–Feynman theorem, 120–1
hybrid orbitals, 63, 142–3
hybridization, 142–6
hydrogen fluoride
  bonding in, 183–5
  population analysis for, 155–6
hydrogen molecule
  configuration interaction for, 60–1
  density-difference distribution for,
    179
  EHF model of, 61, 186–7

hydrogen molecule (*cont.*)
RHF model of, 59–60
hydrogen molecule ion
first excited state of, 178
ground state of, 171–8
hydrogen peroxide, internal rotation in, 192–4
hydrogenation energies, 100, 101

inner-shell binding energy shifts
correlation with atomic charges, 166–8
and Koopmans' theorem, 161–3
potential model of, 164–5
in–out correlation, 186
internal rotation
basis-dependence of barriers to, 189, 193
in ethane, 188–92
in hydrogen peroxide, 192–4
interpair correlation, 65, 71
intramolecular forces, 119–21
intrapair correlation, 65, 71
ionic bonding, 125–6, 182–5
ionization energies, *see* electron binding energies
irreducible representations, 19
eigenfunctions of Hamiltonian as basis for, 23
isomorphic groups, 26

Kekulé structures, 140
Koopmans' theorem, 53–4
and inner-shell binding energy shifts, 161–3

left–right correlation, 186
linear combinations
of atomic orbitals (LCAO), 29, 85–6
method of, 6–9, 81
lithium atom, 42
lithium fluoride
bonding in, 126, 182–3
density-difference distribution for, 125
dipole moment of, 57, 182
population analysis for, 158–9
lithium hydride, 50, 53
bonding in, 182, 183
lithium molecule
density-difference distribution for, 123–5, 181
total charge-density distribution for, 122
localization methods, 131–3
localized orbitals, 129–47
in multiple bonds, 137–8

in $\pi$ electron systems, 138–42
in water, 133–7, 143, 145
transferability of, 145–6
localized pair functions, 146–7

matrix representation
in group theory, 17–20
of Hartree–Fock theory, 80–3
of the Schrödinger equation, 8–9
methane
electron correlation in, 73–4, 147
localized orbitals of, 129
orbital energies of, 54
population analysis for, 156
methylene
geometry of, 102–5
singlet–triplet ($^1A_1$–$^3B_1$) separation of, 105–6
minimal basis, 84, 95–8
molecular geometry, 97–8, 102, 103–5, 108
molecular orbitals (MO)
canonical, 133
Gaussian-type, 88–95
as linear combinations of atomic orbitals, 29, 85–6
Slater-type, 86–8
molecular symmetry, 27–30
Mulliken population analysis, *see* population analysis
multiconfigurational self-consistent field method (MC-SCF), 67
multiple bonds, localization in, 137–8

naphthalene, localized $\pi$ orbitals in, 141–2
natural orbitals, 114–15
net atomic population, 150
$NiF_6^{4-}$, population analysis for, 153–4
nitrogen molecule
density-difference distribution for, 123–5, 181
orbital energies of, 54
total charge-density distribution for, 122
non-empirical theory, 1

one-electron density function, 111–17
one-electron properties, 117–21
expectation values of, 119
open-shell states, 51–3
orbital angular momentum, 13, 30–4
operators for, 31
projection operators for, 32
orbital configuration, 66
orbital density, 111, 126–9

# Index

orbital energies
  definition of, 44
  of inner shells, 159–68
  and Koopmans' theorem, 53–4, 161–3
orbital net atomic population, 150
orbital overlap population, 150
orbital symmetry constraint, 47, 48
  relaxation of, 61, 63
orbitals, complete set of, 50
  see also atomic orbitals; molecular
    orbitals
orthogonality, 3
orthonormal functions, 3, 7
  construction of, 40
overlap densities, 148–9
overlap populations, 149–50

$\pi$ orbitals, localization of, 138–42
pair-correlated wave functions, 70–3
pair-correlation energies, 71–3, 146–7
Pauli principle, 24–7, 115
  and the chemical bond, 179–80
  in the orbital approximation, 38
Pauling wave function, 172
permutation group, 24–7
  antisymmetric representation of, 25–7
  symmetric representation of, 26–7
perturbation theory, 9–12, 56
  and electron correlation, 67–73
phenanthrene, localized $\pi$ orbitals of,
    141–2
photoelectron spectroscopy, 53, 159–63
polarization functions, 86
  and population analysis, 157–8
population analysis, 148–59
  basis-dependence of, 152, 154–9
primitive Gaussians, 94
projected Hartree–Fock theory (PHF),
    58
projection operators
  complete set of, 24
  for construction of symmetry
    orbitals, 29
  definition of, 23
  for orbital angular momentum, 32
  for spin, 35
pyridine, 152
pyrrole, 151, 152

radial correlation, 63
reducible representations, 19
relativistic corrections, 2, 76–9
relaxation
  of double-occupancy constraint, 58,
    61, 63
  of electron distribution, 53, 54, 101

of orbital symmetry constraint, 61, 63
representations, matrix of group, 17–20
  basis for, 18
  characters of, 20–1
  dimensions of, 17
  equivalent, 19
  irreducible, 19
  reducible, 19
restricted Hartree–Fock theory (RHF),
    48–53
  for closed-shell states, 49–51
  definition of, 51–2
  matrix representation of, 80–3
  for open-shell states, 51–3

Schmidt orthogonalization, 40
Schrödinger equation, 1–12
  and group theory, 22–3
secular determinant, 6, 81
secular equations, 6, 8, 81
self-consistent field equations, 82–3
self-consistent field method (SCF), 45,
    82–3
  multiconfigurational, 67
Slater determinants, 38
  complete set of, 52
  invariance to unitary transformation,
    130
  as spin-orbital configurations, 65
Slater functions, see Slater-type orbitals
Slater-type orbitals (STO), 83–5, 86–8
  Gaussian representation of, 96–7
spatial symmetry, 13, 27–34
spin
  angular momentum, 13, 34–6
  coordinate, 25
  eigenfunctions, 34–6
  projection operators for, 35
spin density, 58, 114
spin-orbital configuration, 65
spin-orbitals
  antisymmetrized product of, 37–8
  complete set of, 45
  definition of, 37
  orthogonality of, 39
spin-unrestricted Hartree–Fock theory
    (SUHF), 58
substituted (excited) configurations,
    55–6
symmetry
  electron permutation, 13, 24–7
  molecular, 27–30
  spatial, 13, 27–34
  spin, 13, 34–6
symmetry-adapted wave functions, 23
symmetry operations, 14, 27

204